Cognitive Systems Monographs

Volume 28

About this Series

The Cognitive Systems Monographs (COSMOS) publish new developments and advances in the fields of cognitive systems research, rapidly and informally but with a high quality. The intent is to bridge cognitive brain science and biology with engineering disciplines. It covers all the technical contents, applications, and multidisciplinary aspects of cognitive systems, such as Bionics, System Analysis, System Modelling, System Design, Human Motion, Understanding, Human Activity Understanding, Man-Machine Interaction, Smart and Cognitive Environments, Human and Computer Vision, Neuroinformatics, Humanoids, Biologically motivated systems and artefacts Autonomous Systems, Linguistics, Sports Engineering, Computational Intelligence, Biosignal Processing, or Cognitive Materials as well as the methodologies behind them. Within the scope of the series are monographs, lecture notes, selected contributions from specialized conferences and workshops.

Advisory Board

More information about this series at http://www.springer.com/series/8354

Eris Chinellato · Angel P. del Pobil

The Visual Neuroscience of Robotic Grasping

Achieving Sensorimotor Skills through Dorsal-Ventral Stream Integration

Springer

Eris Chinellato
School of Computing
University of Leeds
Leeds
UK

Angel P. del Pobil
Department of ICC
Jaume I University
Castellón
Spain

ISSN 1867-4925 ISSN 1867-4933 (electronic)
Cognitive Systems Monographs
ISBN 978-3-319-20302-7 ISBN 978-3-319-20303-4 (eBook)
DOI 10.1007/978-3-319-20303-4

Library of Congress Control Number: 2015942181

Springer Cham Heidelberg New York Dordrecht London

Printed on acid-free paper

Springer International Publishing AG Switzerland is part of Springer Science+Business Media (www.springer.com)

To Diego and Lena.
For making everything worth.

Eris Chinellato

To my mother in her 80th birthday.
Por su desvelo incesante, que con el paso de los años no cesa.

Angel P. del Pobil

Preface

Until recently, neuroscience and robotics were utterly unrelated subjects, requiring completely different background, skills and methodologies. Nowadays, the distance between the two fields is being constantly shortened by the progress in computational modeling, and the construction of increasingly skilled autonomous artificial agents inspired by the abilities and behavior of living beings. The astounding discoveries that are being recently achieved by brain scientists constitute the fundamental building blocks for computational neuroscience and biomimetic robotics. This book presents interdisciplinary research which pursues the mutual enrichment of neuroscience and robotics research.

Grasping and manipulation of every kind of object is arguably the most distinctive practical skill of human beings, and erect posture has likely evolved in order to free the upper limbs and make of the hands two unmatchable tools. Despite the great efforts that are being put on it, grasping in robotics is largely an unsolved problem, due to its inherent complexity and the still limited adaptive skills of present day robots in visual and visuomotor behaviors. In our approach, the task of object grasping is dealt with by mimicking, as accurately as possible, the brain mechanisms which underlie planning and execution of grasping actions in humans and other skilled primates.

The principal contribution of the presented research is the definition and implementation of a functional model of the brain areas involved in vision-based grasping actions. The model constitutes a bridge between cognitive science and robotics research, and includes all the steps required for performing a successful grasping action from visual data. The subdivision of visual processing into the dorsal and ventral cortical streams, respectively dedicated to action-oriented and perception-oriented vision, is thoroughly taken into account. Hypotheses regarding the mechanisms that allow to achieve complex interactions with the peripersonal space, through the integration of the data provided by the streams, are put forth. Transfer functions are proposed for modeling the visuomotor transformations performed by the brain areas most critical in grasp planning and execution.

The particular attention payed to the functional role of brain areas makes the model especially suitable for implementation on a real robotic setup, and a full vision-based robotic grasping system has been developed following its guidelines. Visual information regarding an unknown target object is acquired and transformed into a basic representation onto which two concurrent processing mechanisms are performed. The dorsal stream extracts and analyzes possible grasping features, while the ventral stream performs object classification. Dorsal and ventral visual data are merged for estimating shape, size and position of the object, and a grasping plan is devised which takes into account both visual data and proprioceptive information on the state of the arm and hand. Grasp execution is performed with the aid of tactile feedback in order to achieve a stable final hand configuration.

Grasping experiments have been performed on real objects unknown to the system, and the obtained results attest the achievement of our two concurrent goals. On the one hand, the system can safely perform grasping actions on different unmodeled objects, denoting especially reliable visual and visuomotor skills. This confirms that the new research path we propose, according to which robotic grasping can be based on the integration of the two visual processing channels of the primate brain, is significant and worth further exploration. On the other hand, the computational model and the robotic experiments help in validating theories on the mechanisms employed by the brain areas more directly involved in grasping actions. This book offers new insights and research hypotheses regarding such mechanisms, especially for what concerns the interaction between the streams. Moreover, it helps in establishing a common research framework for neuroscientists and roboticists regarding research on brain functions.

London
April 2015

Eris Chinellato
Angel P. del Pobil

Acknowledgments

During the development of this work, we have had the opportunity to discuss and collaborate with outstanding researchers from neuroscience, robotics, and all the shades in between. Among them, we would like to mention Umberto Castiello, Mel Goodale, Jody Culham, Yiannis Demiris, Jose Carmena, Emilia Barakova, Luc Berthouze, Enrico Pagello and Emanuele Menegatti.

A very special thanks goes to the late Prof. José Mira. Pepe sowed in us the seed from which our wonder at the machinery of the brain grew and grew. He was our friend and mentor; an everlasting source of inspiration.

We are deeply indebted will all the members of the Robotic Intelligence Laboratory at Jaume I University (Castellón, Spain), where most of the research work described in this book has been conducted. Particularly valuable has been the help of Beata Grzyb, Antonio Morales, Gabriel Recatalá and Mario Prats.

We would like to express our gratitude to our friends and families, especially to our parents, wives and children, who gave us constant encouragement and tireless support. Double thanks to Oscar Chinellato, who provided important practical contributions to this work.

Finally, the research presented in this book would not have been possible without the financial support received by the authors and the UJI Robotic Intelligence Laboratory from different funding agencies, such as Ministerio de Economía y Competitividad (DPI2011-27846), Generalitat Valenciana (PROMETEOII/2014/028), Universitat Jaume I (P1-1B2014-52) and the European Union through the Seventh Framework Programme (FP7/2007-2013).

Contents

Chapter 1
Introduction

Life sciences and engineering research have been long considered as two unrelated and independent subjects. Researchers of either fields are rarely required to possess more than basic notions regarding the other. Moreover, it is often perceived as pointless for a scientist to delve into themes that are seemingly not directly relevant for her/his research. The implicit assumption is that not much can be learnt from other fields that is useful for the advancement of one's studies. Nevertheless, illustrious exceptions to this way of thinking can be found throughout human history, like Aristotle, Averroes, Leonardo Da Vinci, Gottfried Leibniz, and for the last century John von Neumann, Herbert Simon, Bertrand Russell, just to mention a few universally known names of true interdisciplinary scientists. Compared to the pioneering work of such geniuses, it is nowadays much easier to access all kinds of human knowledge, for anyone that is just curious enough to do it. Also in the scientific and engineering communities the trend is clearly changing, and it is getting more common to find researchers who deal with problems from different points of view provided by complementary disciplines.

Two seemingly unrelated communities that are increasing their mutual interest are neuroscience and robotics. The rapidly growing number and quality of conferences and projects dedicated to robotic applications inspired by neural mechanisms confirm the interest of roboticists in cognitive sciences. On the other hand, the increasing attention of the neuroscience community regarding robotics research is corroborated by the trend in the number of research papers indexed in PubMed that cite the word *robotics*, depicted in Fig. 1.1. The fundamental importance of technological aspects in life science research nowadays partly explains such trend. Another explanation lies in the greatly improved biological plausibility of the newest artificial systems, which are becoming of interest for the people who study the natural systems that inspired them.

Artificial intelligence constitutes the natural bridge between cognitive science and autonomous robotics. The field was born with the idea of emulating human capabilities in problem solving. Amazing advancements have been achieved throughout the years, and control systems based on artificial intelligence techniques are nowadays common in every house. Nevertheless, artificial sensorimotor skills have

E. Chinellato and A.P. del Pobil, *The Visual Neuroscience of Robotic Grasping*,
Cognitive Systems Monographs 28, DOI 10.1007/978-3-319-20303-4_1

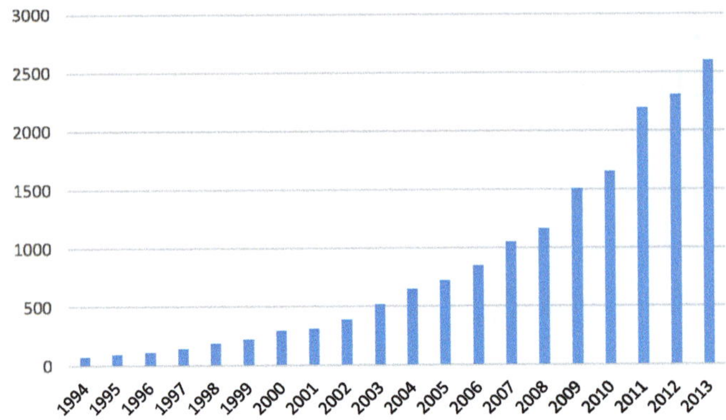

Fig. 1.1 Trend of the number of papers indexed in PubMed which cite the word *robotics* (generated from http://dan.corlan.net/medline-trend.html)

not developed nearly as fast as reasoning ones, due to engineering constraints, but also to the basic differences between brain mechanisms and artificial machines. Computational neurosciences are trying to bridge this gap, providing artificial agents with abilities more deeply inspired by natural mechanisms, in order to achieve, in a medium-long term, better efficiency and more adaptive behaviors.

Robotics often aims at reproducing in artificial beings the most relevant skills characterizing animals in general and humans above all. One of the most distinctive abilities of humans, and to a minor extent of other primates, is the handling of every kind of objects in a dexterous way. Grasping and manipulating skills have been constantly pursued in robotics for their theoretical and practical implications. Nevertheless, in spite of the amount of research and technological efforts, the gap between the prehension performances of primates and that of autonomous robots is still wide, especially in unstructured, real environments. Structural reasons explain such gap. In the first place, technological development in the building of artificial hands and limbs is progressing fast, but the dexterity and versatility of the human hand is still out of reach for the moment. As a consequence, not only technological applications but also modeling efforts are affected, because detailed models of the capabilities of real hands cannot yet be applied to artificial ones.

In addition, the software approach is inherently unsuitable for the emulation of brain functions. The extraordinary skills of the brain are inextricably related to its structure and its tight interconnections with the rest of the body. The distinction between software and hardware simply does not hold if referred to the brain. Still, keeping with the current technology, computational models of brain mechanisms help in two ways. They complement neuroscience findings aiding in the proposal and validation of hypothesis and theories, building on the laws of physics and chemistry, and on fundamental biological concepts such as adaptation through evolution or embodiment. On the other hand, they provide practical insights on how artificial agents, both software and hardware, can be built in order to achieve certain skills.

Following a truly multidisciplinary approach, this book deals with the task of vision-based grasping pursuing two interconnected and complementary goals. The first goal is to obtain a unified schema of the mechanisms and functionality of the brain areas most important for the planning and execution of grasping actions. Implementation of such schema on a real robotic setup allows to verify the appropriateness and plausibility level of the hypothesized mechanisms. As a second goal, those mechanisms endow the robotic system with advanced grasping capabilities not attainable with different approaches. In fact, although biomimetic robotics is a rapidly developing field, the limited literature about biological inspiration in robot grasping at cognitive level suggests that the subject has still much to offer.

Our proposal is based on the two streams theory (Goodale and Milner 1992), according to which visual information in the brain is processed along two main parallel pathways. The pathways evolved for different purposes, being the ventral stream dedicated to perceptual tasks such as object recognition, and the dorsal stream to action-oriented visual tasks, such as motion detection or distance estimation. Nevertheless, recent findings point out the existence of strict relations between them, and the contribution of both pathways is necessary in order to allow proper interaction of human beings with the world, such as in grasping actions.

This book introduces a full framework of the sequence of sensorimotor transformations required to plan and execute hand movements suitable to grasp nearby objects. Present day research on robotic vision-based grasping has been compared with up-to-date neuroscience findings. The proposed framework has been conceived to be applied on a robotic setup, and the analysis of neuroscience findings has been performed taking into account not only biological plausibility, but also practical issues related to implementation constraints. The first contribution of the research presented in this book is thus a model of the brain mechanisms behind vision-based grasping. Such model is faithful to the neuroscience data, but also focused on practical aspects, in order to get a functional understanding of the information flow, and the way it is actually transformed by brain areas, beyond general explanations. The second contribution is a new approach to vision for grasping in robotics, which aims at improving the emulation of human skills through the integration of the information flows proceeding from the two visual pathways.

In the developed model, special attention has been devoted to how visual data are represented and processed at different stages along the dorsal stream. Ventral stream modeling is mainly aimed at extracting information useful for grasping actions, and many details of its processing are not considered. Hand dexterity is obtained joining pure manipulating skills with a visual system especially suitable for estimating the features of nearby objects. In fact, an efficient interaction with the environment requires the ability of estimating distance, size and shape of surrounding objects. Such ability can only be achieved through the use of binocular vision. Not only mammals, but even those birds most skilled in picking up objects (or animals) with their beaks have frontally placed eyes which allow for stereoscopic vision. Sometimes though, stereopsis alone is not enough, and there is the need to complement it with other sorts of visual information, such as motion-related cues, texture, shading and so on. Neuroscience research on primates showed that visual information is indeed processed in a highly parallel way. Different cues for the same stimulus are processed, compared and optimally merged in order to provide increased estimation reliability through redundancy. Cue integration is a major principle in the processing of sensory stimuli, and vision is not an exception.

The first part of the model implements the integration of monocular and binocular retinal information for estimating object pose. The obtained modular computational structure is composed of various stereoscopic and perspective estimators and different merging methods. The framework has been implemented on a real robotic setup at the Robotic Intelligence Lab of Universitat Jaume I, and provides the system with precise and reliable pose estimation capabilities. Experimental results suggest that the principle of cue integration can make robot sensory systems more reliable and robust. The same results show that the model is able to reproduce experimental data obtained in human studies.

The second part of the model is focused on the coding and transformation of visual features into suitable hand shapes. The task to perform is a safe object grasping action, based mainly on visual input. The integration of sensory information of different modalities, visual, tactile and proprioceptive, is complemented with data regarding object identity, and with criteria for action selection. Tactile response is employed in order to adjust the hand configuration to the object and make the grip more stable even in the case of perturbation of the working environment. Again, the approach is coherent with research on primates but also directly suitable for robotic implementation. The obtained artificial grasping behavior is relevant for the robotic community and useful for the interpretation of brain functions.

Summarizing, this book describes the definition and implementation of a model of the extraction and integration of information relevant for grasp planning and execution. The model is both biologically plausible and carefully tailored to be applied to a robotic setup. Experimental results show that, while the robot is provided with advanced grasping skills, effects predicted by neuroscience research are reproduced.

The book is organized as follows. Chapter 2 reviews the neuroscience literature of brain areas involved in grasping actions, following a logical sequence from visual acquisition to action execution, and paying special attention to the possible interactions between the streams. The current approaches to vision-based grasping in robotics are exposed and compared to neuroscience research in Chap. 3. The chapter includes a proposal for grounding sensorimotor interactions to symbolic representations in robotic grasping, in Sect. 3.3. Chapter 4, after a description of related models and a methodological discussion, outlines the modeling proposal, describing briefly the brain areas involved, the basic functions they perform and their connectivity. The model is then detailed in the following two chapters. In Chap. 5 the mechanisms for extracting and merging visual cues to grasping features are described. Chapter 6 presents the transformation from visual input to hand configurations, and the planning and execution of grasping actions. Both chapters provide modeling details and experimental results of the application of the proposed mechanisms to the robotic setup. In Chap. 7, complementary research, alternative interpretations and modeling approaches, and possible future research directions are presented and discussed.

Reference

Goodale MA, Milner AD (1992) Separate visual pathways for perception andaction. Trends in Neurosci 15(1):20–25

Chapter 2
The Neuroscience of Action and Perception

The visual cortex of humans and other primates is composed of two main information pathways, called *ventral stream* and *dorsal stream* in relation to their location in the brain, depicted in Fig. 2.1. The traditional distinction put forth by Ungerleider and Mishkin (1982) and detailed by Goodale and Milner (1992) talks about the ventral "what" and the dorsal "where/how" visual pathways. In fact, the ventral stream is devoted to perceptual analysis of the visual input, such as in recognition, categorization, assessment tasks. The dorsal stream is instead concerned with providing the subject the ability of interacting with its environment in a fast, effective and reliable way. This second stream is directly involved in estimating position, shape and orientation of target objects for reaching and grasping purposes.

The two cortical systems related to the visual streams were previously considered to act nearly independently (Milner and Goodale 1995). However, although recent studies confirm that the dorsal stream is more oriented toward action-based vision, whilst the ventral one is more suitable for categorization, their interaction seems to be extremely important for allowing both of them to function properly (Goodale and Milner 2004).

The tasks performed by the two streams, their duality and interaction, constitute the neuroscientific basis of the research described in this thesis, and this chapter is devoted to a detailed explanation of the related concepts.

2.1 The Two Cortical Streams of Visual Elaboration, Fundamental Roles and Proofs of Dissociation

Looking at an object with grasping purposes activates a neural pathway which is not active when grasping actions are not involved. This activation seems to represent a potential grasping action, and is reinforced when the action is actually performed. Some neurons in the *anterior intraparietal area* (AIP) of posterior parietal cortex in monkeys are found to be active when grasping some particular objects, but also

E. Chinellato and A.P. del Pobil, *The Visual Neuroscience of Robotic Grasping*,
Cognitive Systems Monographs 28, DOI 10.1007/978-3-319-20303-4_2

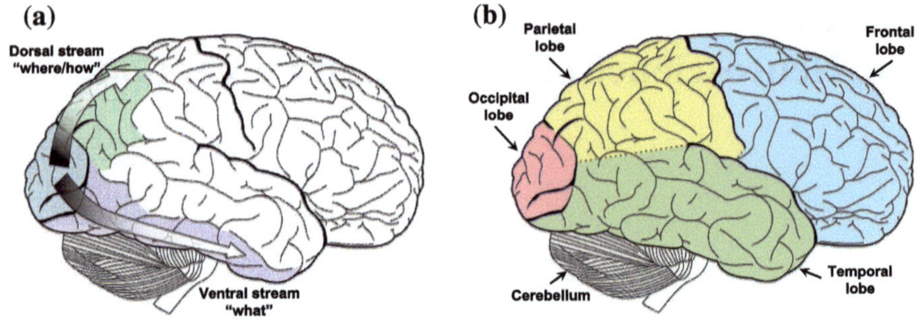

Fig. 2.1 Visual streams and cortical lobes of the human brain. **a** Dorsal and ventral streams. **b** Cortical lobes and cerebellum

when looking at them with the purpose of grasping (Sakata et al. 1998). Some other neurons of the same area are sensitive to the size or orientation of objects, and to hand postures. Area AIP is not activated when the task is to recognize or classify objects and no practical interaction is required. A very similar pattern has been found in humans as well (Culham 2004; Cohen et al. 2009; Hoeren et al. 2013), thanks to *fMRI* and *TMS* research (see boxes below).

fMRI is a non-invasive neuroimaging technique that infers the activity of brain areas from their hemodynamic response. The most common type of fMRI, called BOLD (Blood Oxygen Level Dependent), measures regional differences in oxygenated blood. BOLD-fMRI is by far the most used neuroimaging technique due to its very high spatial and temporal resolution, although the exact relation between neuronal activity and blood oxygenation is still a matter of debate (Culham 2001; Logothetis et al. 2001; Heeger and Ress 2002; Snow et al. 2011).

Transcranial magnetic stimulation (TMS) is a non-invasive brain study technique that creates localized, completely reversible lesions to the brain cortex through the induction of weak currents by rapidly changing magnetic fields (Pascual-Leone et al. 2002). The technique allows to study the functionality and connectivity of brain areas by inactivating them during the execution of certain tasks. The advantage of TMS over imaging methods is that an observed disruption implies the direct involvement of the stimulated area in the tested function. On the other hand, TMS spatial resolution is lower than in fMRI, and the results are strongly dependent on the exact stimulation, in space and time, of the target area.

In the dorsal visual stream of the primate brain, there is thus an area especially dedicated to encode the 3D features of objects in a format suitable to be used for planning and executing grasping actions. Similarly, a large part of the human brain close to the lateral-occipital sulcus (the *lateral-occipital complex*, LOC) is dedicated to recognize visually-presented stimuli, such as objects or faces, but is not directly involved in action execution toward them (Kourtzi and Kanwisher 2001). LOC is probably the most typical ventral stream area.

2.1.1 A Dual Mechanism for Vision

The dualism between "vision for action" and "vision for perception" had been hypothesized long time before neuroimaging research (Goodale et al. 1991; Milner and Goodale 1993). Studies with neurally-impaired people, especially on two categories of brain damages, *visual agnosia* and *optic ataxia* (see box), suggested such dualism.

Visual agnosia (from Greek: *a-gnosis*, lack of knowledge) is the name given to a number of different disorders and syndromes in which visual object recognition is impaired (Farah 2004). Of particular interest for the two streams research is *visual form agnosia*, a type of agnosia that affects identification of shapes even though the subjects have preserved visual acuity, color vision, tactile recognition, and are able to move correctly and properly grasp objects presented in their peripersonal space (Milner et al. 1991; Rice et al. 2006a).

Optic ataxia (from Greek: *a-taxis*, lack of order) occurs when the patient has a deficit in visually-guided arm movements that cannot be explained by motor, somatosensory, or visual acuity deficits (Buxbaum and Branch Coslett 1997; Glover 2003). People affected by optic ataxia are unable to grasp common objects if not very clumsily and unreliably, although their recognition and classification skills are totally spared (Milner and Goodale 1995).

The apparent complementarity of the two impairments have been of great help for the elaboration of the two streams theory. Recent neuroimaging studies revealed that visual agnosia is caused by damages to the LOC and nearby areas, whereas damages to the dorsal stream around AIP provoke optic ataxia. For example, the brain of patient DF, suffering from visual form agnosia, does not show activation related to object identification, because her ventral stream is damaged (James et al. 2003; Whitwell et al. 2014). Nevertheless, she is able to correctly perform grasping actions, and her parietal activation is rather similar to control subjects, including in the anterior intraparietal sulcus during grasping. The opposite behavioral patterns are observed in optic ataxic patients (Goodale and Milner 2004).

Evidence for the different role and processing mechanisms of the two pathways has been provided during the last two decades by plenty of studies following different research approaches and techniques. Recent fMRI research showed the complementary responsiveness of the two streams in identification and spatial analysis of visual stimuli (Valyear et al. 2006; Thoma and Henson 2011). Such dissociation is confirmed for situations in which the action is observed and not directly performed by the subject (Shmuelof and Zohary 2005), or only pantomimed Makuuchi et al. (2012). Considering two of the most representative areas of the streams, AIP for the dorsal, and LOC for the ventral stream, the former shows differential activity during grasping movements with respect to reaching, whilst the latter does not. On the other hand, LOC activates whenever a recognizable object is visible (compared to scrambled images), whilst AIP only when a potentially graspable object is in view (Culham et al. 2003).

Behavioral studies based on optical illusions, distractor stimuli and concurrent tasks suggest that visual information is analyzed and processed differently by the streams (Winkler et al. 2005; Plewan et al. 2012). According to Westwood and Goodale (2003a), explicit object perception in the ventral stream is "scene-based" and the size and location of an object is represented contextually with the size and location of nearby objects. The control of object-directed actions by the dorsal stream follows instead an "actor-based" frame of reference, in which object location and size are represented with respect to the subject body, and especially to hand and arm. Dorsal visual analysis is driven by the absolute dimensions of the target object, and other objects in the environment are likely to be considered and hence taken into account only as potential obstacles (Ansuini et al. 2007b). Another distinction talks about holistic and analytical visual representations (Ganel and Goodale 2003; Red et al. 2012): object dimensions that are perceived globally by the ventral stream are, in the same situation, processed locally by the dorsal stream if a visually-guided action is directed at the object.

Several studies (see e.g. Grill-Spector et al. 1999; James et al. 2002) demonstrated that ventral stream areas such as LOC show adaptation for different views of the same object, denoting viewpoint invariance. On the contrary, areas of the intraparietal sulcus do not exhibit such invariance, and respond to different views as they were different objects. This suggests a more "pragmatic", action-oriented on-line processing along the dorsal stream, focused on the actual situation of the environment rather than on objects' implicit quality. Even access to memory seems to be different for the two streams, and working memory related to spatial location and visual appearance is probably located in different subsystems (Darling et al. 2006; Kawasaki et al. 2008).

The streams dissociation has thus been confirmed, but also criticized, by the neuroscientific community, and the original theory is constantly being revised and updated. The trend is toward a more integrated view of the functioning of the two streams, that have in many cases complementary tasks (Goodale and Westwood 2004; Zachariou et al. 2014).

2.1.2 *Brain Pathways for Vision-Based Grasping*

The anatomy of the visual and motor cortices of human and closer superior primates is well known. Although the knowledge regarding associative regions of the brain, such as the posterior parietal or the inferior temporal cortices, is less established, it is possible to outline a simplified schema of the brain areas more directly involved in vision-based grasping actions. Those areas more thoroughly considered in this work are depicted in Fig. 2.2. A longer list of brain areas, with acronyms or short names and references to the sections in which they are described is provided in Table 2.1. Here, only an overview of the two pathways is given, and more details are provided in the rest of this chapter.

Visual data in primates flows from the retina to the lateral geniculate nucleus (LGN) of the thalamus, and then mainly to the primary visual cortex (V1) in the occipital lobe. The two main visual pathways go from V1 and the neighbor area V2 to the posterior parietal cortex (PPC) and the inferior temporal (IT) cortex. Through the dorsal pathway, object related visual information flows through area V3A and the caudal intraparietal area (CIP), which extracts action-related spatial visual properties of objects. Visual data then reaches the anterior intraparietal sulcus (AIP), in which visual features are analyzed in order to plan and monitor the execution of suitable grasping actions. Area AIP projects mainly to the ventral premotor area (PMv), that selects and composes motor primitives (see box on motor primitives) to form complete grasping actions, which execution signals are released by the primary motor cortex (M1).

Object information flowing through the ventral pathway passes through V3 and V4 to the lateral occipital complex (LOC), that is in charge of object recognition. According to recent interpretations of the two streams hypothesis (Goodale and Milner 2004; Ellison and Cowey 2009), the LOC itself is implied in some action-related processing, although the way the two streams communicate is still mostly unknown.

The neuroscience concepts most relevant for the current research are described in the next sections, separated in early visual processing, dorsal, and ventral stream areas. Exhaustive reviews of grasp-related research are Castiello (2005) and Culham et al. (2006). For details regarding visual areas, fundamental studies are Felleman and Essen (1991) and Chalupa and Werner (2003). Most brain regions cited in the text can be localized in Fig. 2.2.

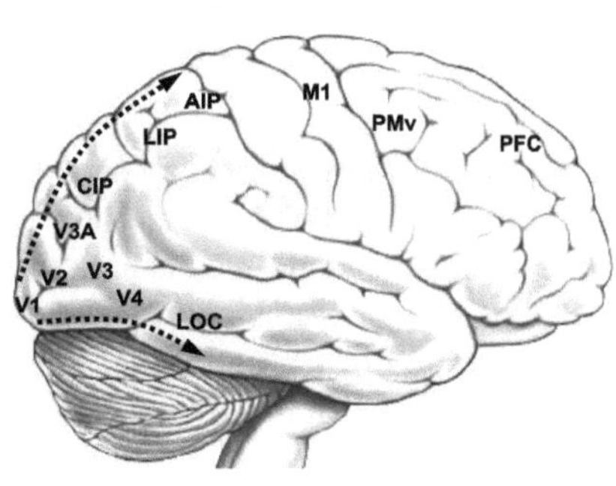

Fig. 2.2 Brain areas involved in vision-based grasping actions

Table 2.1 Principal cited brain areas, with acronym and reference to the section in which they are described

Brain area	Acronym	Sections
Visual areas		
Primary visual cortex	V1	2.2
Visual area 2	V2	2.2
Visual area 3	V3	2.2
Middle-temporal area	MT/V5	2.2
Dorsal stream areas		
V3 Accessory area	V3A	2.2
Intraparietal sulcus	IPL	2.3
Caudal intraparietal sulcus	CIP	2.3.1
Anterior intraparietal sulcus	AIP	2.3.2
Lateral intraparietal sulcus	LIP	2.3.4
Ventral intraparietal sulcus	VIP	2.3.4
Parietal reach region	PRR	2.3.4
Ventral stream areas		
Visual area 4	V4	2.4
Lateral occipital complex	LOC	2.4.1
Ventral occipital temporal area	vTO	2.4.1
Lateral occipital cortex	LO	2.4.1
Motor areas		
Primary motor cortex	M1	2.3.3
Ventral premotor cortex	PMv/F5	2.3.3
Dorsal premotor cortex	PMd	2.3.3
Other areas and structures		
Lateral geniculate nucleus	LGN	2.2
Posterior parietal cortex	PPC	2.3
Somatosensory cortex	SI/SII	2.4.2
Prefrontal cortex	PFC	2.4.2
Basal ganglia		2.4.2
Cerebellum		2.4.2

Differences between humans and other primates are discussed in the text

2.2 Visual Areas and Stream Separation

The *retina* is the visual receptor of the human body. Visual information gathered by the retina is sent through ganglion cells to the *lateral geniculate nucleus* (LGN) of the thalamus. Ganglion cells are of two types: *parvocellular* (P) and *magnocellular* (M); the former are smaller, slower and carry many details such as color, the latter are larger and faster, and rather rough in their representations. Although these two types of cells seem to correspond nicely to the ventral and dorsal stream distinction,

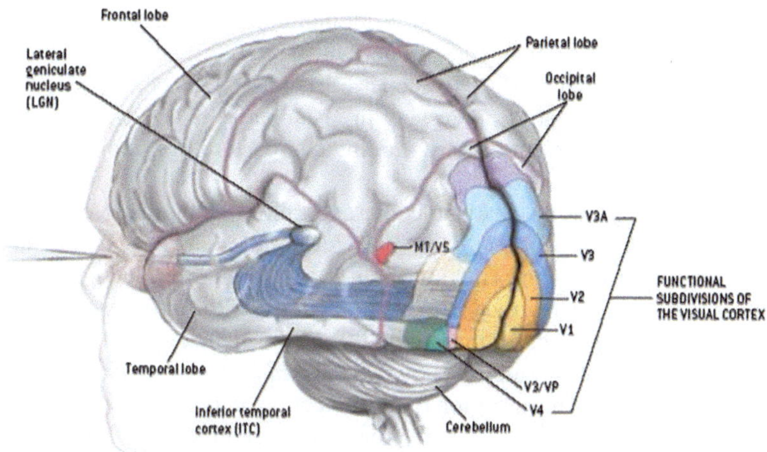

Fig. 2.3 Visual areas in the brain, adapted from Logothetis (1999)

evidence is clearly against a simple correspondence between the subcortical and the cortical pathways, and M and P signals mix largely inside V1 (Maunsell 1992; Ferrera et al. 1992).

The LGN performs a first processing of the visual data and forwards them almost entirely to the *primary visual cortex* (V1) in the occipital lobe (Lee 2003). The primary visual cortex and neighbor visual areas can be localized in Fig. 2.3.

Area V1, also called the *striate cortex*, is organized in a retinotopic manner, respecting the topological distribution of stimuli on the retina. In V1 basic visual features such as colors, bars or edges and their orientation are detected. Visual areas downstream from V1 are called *extrastriate*. The first extrastriate area, V2, receives most of V1 output and projects mainly to visual areas V3 and V4. Area V2 is retinotopic, has receptive fields that are larger than V1's and realizes a matching of V1 features in order to perform moderately complex visual tasks, such as detecting spatial frequencies and textures or separating foreground from background. Visual area V3 is still retinotopic and elaborates on the job of V2 to generate more complex invariant representations. V3 has large receptive fields and ability to detect more complex features regarding orientation, motion, depth and color of stimuli (Gegenfurtner et al. 1997; Adams and Zeki 2001). In V3 the data stream splits into the two pathways: dorsally towards the posterior parietal cortex (PPC) and ventrally to the inferior temporal (IT) cortex. Comparative studies between human and monkey (usually macaques) visual cortices reveal that their brains differ mostly in higher-order cortical regions, downstream from V3-V3A, and are more similar in lowest areas, such as V1 and V2 (Van Essen et al. 2001; Tootell et al. 2003; Tsao et al. 2003).

For what concerns stereoptic processing, binocular disparities (see box on disparities) are present in all visual areas, starting from V1 (Poggio et al. 1988; Cumming and DeAngelis 2001; Parker 2004). Areas V2 and V3 are increasingly capable of depth processing, in accordance with the size of their receptive fields (Backus et al.

2001; Rutschmann and Greenlee 2004). Both in humans and monkeys, area V3A is specialized for stereoptic depth, and computes also relative disparities between pairs of visual stimuli (Tootell et al. 1997; Backus et al. 2001; Tsao et al. 2003). Evidence regarding the role of disparity processing in visual areas is not conclusive though, as the distribution of different disparity tuning curves is rather smooth across areas (Adams and Zeki 2001).

Links between various disparity-selective cells allow to obtain more sophisticated response properties. For example, selectivity for absolute distance is obtained from disparities using additional information about eye position. Computation of disparity gradients is very likely performed in V3A and CIP using the outputs of many simple disparity selective cells (Adams and Zeki 2001; Tsutsui et al. 2005).

Visual area V5, more commonly known as the *middle temporal area* (MT), is very likely the most important brain region for the detection of moving visual stimuli. Both in humans and monkeys MT is selective for speed and direction of moving features (Orban et al. 2003), and its responsiveness to stereopsis signals suggests that it codes also for changes in object orientation (DeAngelis et al. 1998; Nguyenkim and DeAngelis 2003). Even though the contribution of MT is required for performing grasping actions toward moving targets (Schenk et al. 2005), there is no evidence for an involvement of MT in detailed 3D analysis of objects, and thus of its relevance for grasping actions toward static targets (Backus et al. 2001; Tsao et al. 2003).

Most projections in the visual cortex follow the described processing sequence, from V1 to higher order areas, but backprojections are widespread, and visual elaboration in the early visual areas is subject to global context influence, task requirements, and to higher order perception (Lee 2003). According to the modern view, the early visual cortex does not only perform the first, simple stages of visual processing, but is also involved in many higher levels of visual elaboration (Vanni et al. 2004). The visual cortex has been found to be more temporally compact than expected, and transmission times between areas spatially and hierarchically distant are very low (Bullier 2001). Primary areas are thus constantly involved in all stages of visual processing, and higher areas such as MT are in a position to modulate the response of V1 and V2 neurons and suit their response to the requirement of the visual task in a reactive way. Even the LGN seems to be integrated with higher areas, and a short-cut channel of M cells between LGN and MT might be the instrument used by the dorsal stream to quickly separate objects from background, and bias the processing in V1-V2 (Bullier 2001).

2.3 The Action-Oriented Dorsal Stream

Visual region V3A can be considered as pertaining to the dorsal stream, which continues in the posterior parietal cortex, toward the top and sides of the brain. The *posterior parietal cortex* (PPC) is largely recognized as the main associative area of the brain dedicated to the coordination between sensory information and motor response (Sakata et al. 1997). The *intraparietal sulcus* (IPS) separates the superior and inferior lobes of the PPC. Several areas within and close to the IPS are dedicated to different visuomotor tasks (Sereno et al. 2014). Some of these areas are described in this chapter, but special focus is put on its most posterior and anterior sections, CIP and AIP respectively.

Many of the findings explained below concern monkey data, as single cell studies allowed to collect a great deal of evidence regarding the role of intraparietal areas in macaques. Only recently, although in an ever-growing fashion, brain imaging and TMS studies (see box) began to clarify the structure and tasks of the posterior parietal areas in humans.

Differences in control strategies depend also on structure, morphology and kinematics of body and limbs, and it is therefore very difficult to draw a full interspecies parallel (Christel and Billard 2002). The current evidence suggests that the human intraparietal cortex is more complex, and contains visuospatial processing areas that are not present, or much reduced, in monkeys (Grefkes and Fink 2005; Orban et al. 2006a). Neuroscientists argue that, under evolutionary pressure, parietal but not earlier regions adapted to endow humans with specific abilities (Striem-Amit et al. 2012), such as an improved motion-dependent 3D vision for tool manipulation (Vanduffel et al. 2002; Orban et al. 2006a).

Despite the differences, a rather clear parallel between monkey and human AIP is established (Grefkes et al. 2002; Choi et al. 2006; Orban and Caruana 2014). Also, many fundamental connections correspond across species, such as the *anterior intraparietal—ventral premotor* and the *medial intraparietal—superior colliculus* links (Rushworth et al. 2006). Hence, it is a common procedure to consider data of similar species in order to try and work out the mechanisms behind vision-based grasping in humans (Rizzolatti and Luppino 2001; Grefkes and Fink 2005). Important interspecies differences are nevertheless taken into account and discussed in the following sections.

2.3.1 Posterior Intraparietal Sulcus

The most posterior part of the IPS is the *caudal intraparietal sulcus* CIP, which is also referred to as cIPS, pIPS, PI or hCIP in the human case. Area CIP is mainly dedicated to local 3D shape and orientation processing. It receives projections from visual area V3 and V3A and is also active during visually guided grasping.

Neurons in CIP are strongly selective for the orientation of visual stimuli. Two exhaustive studies (Taira et al. 2000; Tsutsui et al. 2001) showed that selectivity toward disparity based orientation cues is predominant, but many neurons also respond (some exclusively) to perspective based orientation cues. Indeed, it seems that cue integration for obtaining better estimates of orientation is performed in this area (Welchman et al. 2005). This sort of processing by CIP neurons is the logical continuation of the simpler orientation responsiveness found in V3 and V3A. Similarly to V3A, CIP is not concerned with general purpose scene segmentation, but rather with processing the 3D layout of target local features (Tsao et al. 2003; Tsutsui et al. 2005). In CIP, orientation of features is represented in a viewer-centered way, so that the coding is especially suitable for visuomotor transformations for reaching-grasping purposes, rather than for feature integration with the purpose of composing complex scene interpretations (Sakata et al. 2005). This is consistent with the position of CIP in a central stage of the dorsal stream. As a further proof of this, CIP does not recognize the same object seen from two different viewpoints (James et al. 2002).

Neurons in CIP have been found to maintain a short-term memory of 3D surface orientation (Tsutsui et al. 2003). This suggests a possible role of CIP in visual tracking and feature matching processes. For example, they might maintain memory of surfaces during active vision, for tracking suitable grasping surfaces.

2.3.1.1 Surface Orientation Selective and Axis Orientation Selective Neurons

Two main neuronal populations have been distinguished in CIP: surface orientation selective and axis orientation selective neurons. *Surface orientation selective* (SOS) neurons were first studied, and their responsiveness described, by Shikata et al. (1996). They respond to a 2D shape in different orientations, but extract the signal of 3D surface orientation from a 2D contour viewed in a linear perspective: i.e., these neurons interpret the stimuli as the silhouette of a square plate slanted in depth (Sakata et al. 2005). Experiments executed changing the proportions of the visual features showed that the responsiveness is maximum for "square" shapes, in which the two major dimensions are similar, and elongation in either width or length inhibits the response. Regarding the third, minor dimension, it seems not affecting the response up to a certain thickness, but if this threshold is overcome a clear decrease in responsiveness can be noted.

The second class of CIP neurons, *axis orientation selective* (AOS) neurons, represent the 3D orientation of the longitudinal axes of elongated objects. According to Sakata et al. (1998), their response increases with decreasing thickness (the two minor dimensions) and with increasing length (the major dimension), showing complementarity with SOS neurons. It is not clear from the provided data if the reduced responsiveness with thicker objects is only due to the relative proportion between the object dimensions or also by some comparison with the hand size. This issue will be discussed in Sect. 6.1.

Some AOS neurons are shape selective, and distinguish for example between cylinders and square columns of similar length and thickness. This suggests that disparity gradients are used in CIP to detect also the curvature of objects (Katsuyama et al. 2005; Naganuma et al. 2005; Sakata et al. 2005). Shape-selective AOS neurons in CIP are thus likely to maintain a prototype of 3D shape representation, as all curved surfaces can be characterized by a shape index and a curvedness index (de Vries et al. 1994).

2.3.1.2 Human CIP

The correspondence between monkey CIP and areas of the human intraparietal sulcus is still problematic, especially if compared with the rather well accepted interspecies matching of early visual and anterior intraparietal areas. Neuroimaging research showed nevertheless that a posterior region of the IPS activates for stimuli similar to those processed by CIP in monkeys, although human CIP seems to be located more medially in the human intraparietal sulcus than in monkeys, as stressed by Grefkes and Fink (2005). An area located in the posterior part of IPS and clearly involved in complex orientation discrimination and coding of 3D object features have been observed by Tsao et al. (2003). The authors call it *caudal parietal disparity region* (CPDR), and suggest that it might be part of the human correspondent of CIP. A similar responsiveness to stereopsis defined stimuli has been registered in other studies (Shikata et al. 2003; Brouwer et al. 2005). Activation in the posterior part of the human IPS has been found using fMRI during orientation discrimination tasks, for both monocular and binocular stimuli (Shikata et al. 2001; Naganuma et al. 2005). Although a clear correspondence is yet to be achieved, the data collected by Shikata et al. (2001, 2003) and other studies clearly indicate that, similarly to its role in macaques, the function of human CIP is that of coding 3D features of target objects for providing AIP with the information necessary for visually-guided hand movements.

2.3.1.3 CIP as a First Meeting Place for the Two Streams

Some findings (see e.g. the work of Tsutsui et al. 2003) suggest that the role of CIP might be more complex than just extracting object visual data and forwarding it to AIP. Memory related activity of CIP neurons indicates that this area might be involved in higher-order 3D visual perception. For example, visual areas V1/V4 do not have such sustained activity, whilst higher ventral stream area LOC has. There are also cues regarding direct connections between CIP and ventral stream areas. Firstly, CIP probably receives input from V4 (Baizer et al. 1991), and this would be the first connection between ventral and dorsal pathways after the splitting (see also Zanon et al. 2010). Moreover, some LOC neurons are selective for orientation and curvature of surfaces, but LOC receives most input from area V4, which is not sensitive to curved surfaces (Orban et al. 2006b). A link between CIP and LOC is the most likely explanation for such findings (Tsutsui et al. 2003). The first link, in the ventral → dorsal direction, could represent a ventral contribution to the process of pose and shape estimation in CIP. In fact, shape recognition allows to follow basic assumptions about objects' geometry and exploit common knowledge about the world in order to estimate size and pose of objects (e.g. to know that an object has square faces permits to use perspective in order to estimate its orientation). The dorsal → ventral link might instead accelerate object identification providing LOC with precise geometric information of local object features. Overall, 3D shape processing seems to possess a contextual nature, and on-line information is probably integrated with abstract representations in order to obtain the most likely interpretations of the visual data (Todd 2004).

Summarizing, CIP has a very precise 3D orientation response, probably obtained through the integration of disparity based, stereoptic cues (prevalent) and monocular, perspective cues. Overall, a population of mixed CIP neurons, including different types of SOS and AOS neurons, is able to provide full information about 3D proportion and orientation of a target shape. This information is forwarded to AIP, where 3D orientation and shape can be coded as a unique, combined feature, and possible *affordances* (see box below) can be generated.

The original definition of *affordance* was introduced to indicate any possible action that an agent can perform in the environment (Gibson 1979). An affordance is related to an object, or to a set of objects, and to the agent abilities. A grasping affordance exists only if a graspable object is present in the environment and if the agent is actually able to grasp it. In this book the term affordance is used with a restricted meaning, to refer to a grasping possibility offered by an object to the human or the robot hand.

2.3.2 Anterior Intraparietal Sulcus

The most frontal part of the IPS is the *anterior intraparietal sulcus*, AIP, sometimes called aIPS or, for humans, hAIP. Both for monkeys and humans, AIP is largely recognized as the area of the brain dedicated to the visuomotor transformations necessary to map visual stimuli onto hand configurations suitable for grasping target objects.

Several electrophysiological studies on macaques monkeys showed that AIP activates at the visualization of a possible target object, and remains active during preshaping and manipulation (Taira et al. 1990; Sakata et al. 1995; Murata et al. 2000). On the contrary, AIP is not explicitly involved in spatial analysis that is not related to action: e.g., it is not active during perceptual size discrimination, for 2D pictures, or for non-graspable objects.

Different AIP neurons are tuned to different objects, to different views of the same object, and to different grips. Although some AIP neurons are specific to one spatial aspect only, similarly to CIP's, axis orientation and shape are often represented as a combined 3D feature in AIP, an probably constitute the full coding of a graspable feature (Sakata et al. 2005). Moreover, some neurons in area AIP discriminate not only between simple solid shapes, but also between complex objects composed of two or more components. According to Sakata et al. (1999), these neurons may be sensitive to very small details critical for the selection of a grip pattern.

In Murata et al. (2000) a detailed description of experiments performed with several different conditions is provided. Neurons in AIP are found to be selectively activated according to shape (one or more of a set including ring, plate, cube, cylinder, cone and sphere), size and orientation of stimuli. Different activation patterns were observed during fixation and visually-guided grasping tasks. Again, selectivity for shape/size/orientation is often merged in a combined selectivity that can be identified as a grasp configuration.

2.3.2.1 Classification of AIP Neurons

Although AIP keeps active from object observation to the end of movement execution, some AIP neurons are selective for one of the following grasping sub-phases: set, preshape, enclose, hold, release (Ro et al. 2000; Debowy et al. 2001). This subdivision is much clearer though in the premotor cortex.

A better documented classification of AIP neurons in subpopulations can be done according to their preferential response in different acting conditions (Sakata et al. 1995; Murata et al. 2000). Three main types of AIP neurons were first classified, *visual* (V), *visuomotor* (VM) and *motor* (M), and the first two classes have been further subdivided into two, *object* (O) and *non-object* (NO), for a total of five neuronal classes:

object type visual-dominant neurons, O-V, respond equally to simple visual presentation (fixation) of graspable 3D objects and during visually-guided grasping actions; these neurons show no activity during grasping in the dark or when direct view of the ongoing action is unavailable;

non-object type visual-dominant neurons, NO-V, respond during visually-guided grasping only, and their activation starts just before hand-object contact; they show no activation during fixation and grasping in the dark;

object type visuomotor neurons, O-VM, are selective during fixation and during grasping actions both in the light and in the dark, but show a clear preference for visually-guided actions compared to fixation and grasping in the dark;

non-object type visuomotor neurons, NO-VM, are selective for grasping actions both in the light and in the dark, with a clear preference for visually-guided actions compared to grasping in the dark; they show no activation during fixation;

motor-dominant neurons, M, are equally responsive during grasping in the light and in the dark, showing no preferential activation between the two cases; they show no activation during fixation.

On a temporal scale, object type neurons, both O-V and O-VM start their activation at the sight of the target object, and seem thus to be in charge of planning the action, transforming the spatial visual information coming from CIP into a more purely grasp-related form. At the action onset and until the contact, visuomotor neurons, both O-VM and NO-VM, reach the top of their activity, revealing a crucial role in the execution of the hand preshaping movement. Neurons NO-V and M also increase their activity during action execution, NO-V neurons only if vision is available, M neurons also in the dark. All five types of neurons remain active during the hold phase until object release, but all of them show a gradually decreasing activation. In the dark, only VM and M neurons stay active, and for the first the responsiveness is reduced compared to the light condition. Hence, neurons in AIP are not only dedicated to plan and begin grasping actions, but also to monitor them during their evolution.

Regarding neural coding, it seems that object type neurons (O-V and O-VM) describe a shape-based representation of objects, whilst motor neurons (M) code for the hand configuration suitable for grasping. Non-object type neurons (NO-V and NO-VM) maintain a someway intermediate representation. Summarizing, it appears that the classification object/non-object/motor accounts for the transformations required to pass from a visual to a motor representation of the target object. The traditional visual/visuomotor/motor classification is more likely related to temporal aspects of action execution.

2.3.2.2 Human AIP

In humans, AIP is located at the junction between anterior IPS and inferior PCS (postcentral sulcus). Again, it is considered the most important area involved in the planning and monitoring of grasping actions. The coincidence with monkey AIP is rather uncontroversial, as many fMRI studies have been consistently showing grasping-related activation in the anterior part of the IPS for more than a decade (Binkofski et al. 1998; Culham et al. 2003; Cavina-Pratesi et al. 2007a). For detailed reviews of such studies, refer to Castiello and Begliomini (2008) and Tunik et al. (2007).

The most relevant difference between species is likely the absence of tactile response in macaque AIP (Murata et al. 2000), contrasted to the clear responsiveness during haptic exploration and purposive manipulation for human AIP (Jäncke et al. 2001; Grefkes and Fink 2005). In fact, AIP is increasingly activated during multimodal processing, suggesting that it might play a specific role in cross-modal transformations of object representation between visual and tactile modalities during grasping (Grefkes et al. 2002). Indeed, the current view of AIP as an associative "visual" area is probably biased by the amount of research on vision, and AIP might finally reveal itself to be as much tactile as visual (Roland et al. 1998).

In humans, AIP is preferentially activated during grasping with precision grips in comparison with full-hand power grips, suggesting a fundamental role in the fine calibration of finger positioning, as required in precision grip tasks (Ehrsson et al. 2000; Begliomini et al. 2007; Cavina-Pratesi et al. 2007b). AIP is probably involved also in controlling action execution by monitoring the difference between an efference copy of the motor command and visual and tactile sensory experience (Rice et al. 2006b). Various studies assign a more dynamic role to AIP beyond grasp planning. For example, transcranial magnetic stimulation of AIP ends in a clear disruption of online grasp control (Glover et al. 2005; Tunik et al. 2005), suggesting that the job of AIP is critical in the online monitoring/adjustment of hand movements.

There are also insights that AIP may execute more "cognitive" tasks and be connected to ventral stream regions. First of all, AIP and nearby areas respond to action recognition when grasping is involved (Fogassi et al. 2005; Shmuelof and Zohary 2005), indicating a more perceptual role than traditionally thought (Culham and Valyear 2006). Also, a region close to the intraparietal sulcus has been found active during object recognition from non-canonical viewpoints (Sugio et al. 1999). The authors suggest that recognition in those cases may be supported by information regarding functional properties of the object, extracted in the CIP-AIP circuit.

Other studies mention direct connections from the inferior temporal cortex to AIP (Fogassi and Luppino 2005; Borra et al. 2007) and other areas of the IPL, suggesting that AIP could use some ventral information in order to plan and execute appropriate grasping actions. Thus, after CIP, also AIP is probably connected, maybe even bidirectionally, to ventral areas, confirming the view that the collaboration between the streams is more strict than previously thought.

2.3.3 Ventral Premotor Cortex (PMv) and Other Motor Areas

The motor cortex occupies the posterior half of the frontal cortex. It is composed of anterior and posterior motor areas, the former connected to the prefrontal cortex, the latter to the posterior parietal cortex. Posterior motor areas can be further subdivided in the *primary motor cortex* M1 (also called F1), and premotor areas F2-F5. Area M1, upon reception of signals coming from the premotor cortex, activates and controls movements of specific body parts. The primary motor cortex is closely linked to corresponding areas in the primary somatosensory cortex S1, which in the case of grasping provides the tactile feedback necessary to adapt the grip to the inertial forces and the object structure (Rizzolatti and Luppino 2001).

A modern view of the organization, function and connectivity of the motor cortex has been proposed by Rizzolatti et al. (1998). The main concept is that the motor cortex is formed by a mosaic of separate areas containing independent body movement representations, which are used in motor control according to the requirements specified by corresponding areas of the posterior parietal cortex (Luppino and Rizzolatti 2000). Thus, parieto-premotor connections form a series of circuits devoted to specific sensorimotor transformations. Rizzolatti et al. (1998) define these circuits as the basic functional units of the motor system, which transform sensory information into action. According to the authors, although their hypothesis is mostly derived from monkey data, brain-imaging and anatomical evidence suggest that the same principles underlie the organization of the human motor cortex as well.

Two such circuits that have been clearly identified in monkeys connect ventral premotor areas F4 and F5 with intraparietal areas VIP and AIP respectively. The former circuit performs the sensorimotor transformations necessary for arm, neck and face movements, the latter permits the execution of hand and mouth movements, and is directly responsible for grasping actions (Luppino et al. 1999). Although their tasks are clearly related, these two circuits are described as anatomically segregated, suggesting a parallel processing between reaching and grasping actions, which are integrated only in the initial planning and the final execution phase, not in the intermediate sensorimotor transformation steps. In humans, kinematic and lesion studies support a parallel and concurrent parieto-premotor processing for reaching and grasping movements (Jeannerod 1997).

The literature description of F5 is consistent with its direct link with AIP. About half of F5 neurons can be considered visuomotor, as their activation begins during object fixation. Although their responsiveness is very similar to the visuomotor neurons of AIP (Murata et al. 1997, 2000), motor specificity of F5 neurons does not depend on the object shape but on the grip used to grasp the object (Raos et al. 2006). Additionally, neurons in F5 can code for full actions, such as grasping or pulling, and for action segments, such as preshaping or holding. Moreover, many F5 neurons are selective for one of precision grip (predominant), finger prehension or whole hand prehension (Rizzolatti et al. 1988). According to Rizzolatti and Luppino (2001), neurons in F5 code for spatial characteristics and temporal segments of grasping movements, hence constituting a vocabulary of motor prototypes to select and compose in the final action.

One of the most popularly known neuroscientific discoveries of the last decades concerns F5. This region is in fact the place where *mirror neurons* were observed for the first time (Di Pellegrino et al. 1992; Rizzolatti et al. 1996). Mirror neurons fire when the subject is performing a certain action, as normal premotor neurons of the same area, but also when the subject observes someone else performing the same action. They have been related to the ability of social interactions through understanding/prediction of other people's movements (Rizzolatti and Arbib 1998), to learning by imitation, and to the explanation of social behavior impairments as in autism (Williams et al. 2001).

The *ventral premotor cortex* (PMv) is still poorly characterized in humans. In fact, fMRI research failed to show consistent activation in the putative human equivalent of F5 during grasping movements (Castiello and Begliomini 2008). Nevertheless, PMv is still believed to play a key role in the preparation and execution of grasping actions. TMS studies showed its importance for grasping, and its task sharing with the *dorsal premotor cortex* (PMd) (Davare et al. 2006). Although the homology with macaque F5 remains controversial, there is a distinct evidence for a dissociation between PMv and PMd roles in controlling precision grasping in humans. Similarly to F5, PMv seems to perform the visuomotor transformations necessary to shape the hand to a target grip (Chao and Martin 2000), whilst PMd may control the correct timing of the action (Davare et al. 2006).

2.3.4 Other Dorsal Stream Areas

Consistently with the view of Sakata et al. (1997); Rizzolatti et al. (1998) and other studies (e.g. Buneo and Andersen (2006)), premotor-parietal circuits perform both direct and inverse coordinate transformation between vision and effector systems, to allow programming and monitoring of complex motor actions. The circuit linking AIP with F5 is not the only one necessary for the execution of accurate grasping actions, as proximal limb movements, eye and head coordination, and various posture movements are all required in order to allow the hand to perform a correct shaping sequence. Some other well-recognized areas of the posterior parietal cortex are briefly described below (for more detailed descriptions please refer to Culham et al. (2006) and Grefkes and Fink (2005)).

LIP is the *lateral intraparietal area*, also called *parietal eye field*, PEF and, for humans, hLIP or hPEF. Evidence for the role of this region in humans, and inter-species analogies are well recognized, although most studies indicate a more medial location of human LIP compared to monkey LIP. Area LIP aids in the execution of saccadic eye movements and transformation between retinotopic and head-centered coordinates (Grefkes et al. 2004; Scherberger and Anderson 2004; Sereno et al. 2014). Neurons in LIP have been found to be modulated by both proprioceptive and retinal stimuli, suggesting that LIP contributes to distance estimation combining vergence and disparity through a gain modulation effect (Naganuma et al. 2005; Genovesio and Ferraina 2004).

VIP is the *ventral intraparietal area*, involved in head movements coordination and near-head space analysis. Area VIP receives strong input from the motion selective area MT and responds to optic flow, detecting movements in head-center coordinates during self-motion (Bremmer et al. 2001). It is also likely that VIP contributes to multimodal integration in the dorsal stream, as it is activated by visual, tactile and auditory stimuli, showing congruent receptive fields across modalities (Bremmer et al. 2002; Lewis and Essen 2000). Although many of its properties have been observed in human cortical areas, consistent data are not yet available for defining a clear human correspondent of monkey VIP.

PRR is the *parietal reach region*, the area of the PPC dedicated to perform the reference-frame transformations and the sensorimotor coordinations necessary for pointing and reaching movements (Grefkes et al. 2004). Most probably, PRR performs also a monitoring of ongoing actions and adjust them according to an efference copy of the motor signal (Kalaska et al. 2003; Gréa et al. 2002). In monkeys this area is quite well circumscribed, it includes MIP—the *medial intraparietal sulcus*—and visual area V6A. On the other hand, pointing and reaching movements in humans seem to involve several disjunct areas of the superior parietal lobe, such as: V6A (Pitzalis et al. 2013), MIP/mIPS (medial intraparietal sulcus, Reichenbach et al. (2014)), PCu (precuneus) and POJ (parieto-occipital junction); the exact purpose of the movement (reaching vs. pointing) and the position of the target (central vs. peripheral vision) are among the factors that differentiate the cortical activation (Grefkes et al. 2004; Davare et al. 2012).

2.4 Object Recognition and Stream Integration

Compared to early visual areas, neurons in higher order areas such as V4 and the LOC have larger receptive fields, and can integrate information across long distances in the visual field (Grill-Spector et al. 1998; Bullier 2001). Object areas along the ventral stream can thus represent visual stimuli with increasingly complex and invariant representations.

2.4.1 The Lateral Occipital Complex

The *lateral occipital complex* (LOC) is the region of the human brain in which viewpoint invariant object representation for immediate visual recognition is performed (Grill-Spector et al. 2001). The LOC receives high level visual input from V4 and integrates visual elements that share similar attributes of orientation, color, or depth into objects and extract them from the background (Grill-Spector 2003). Object representation in LOC is highly invariant with respect to the stimulus type, showing equally good performances with either 3D or silhouette images, different color maps, lightning and so on. This suggests a higher level, conceptual representation of objects, independent of the actual stimulus that allowed recognition (Kourtzi and Kanwisher 2000).

The LOC is constituted by two different areas, anterior and posterior, which seem to maintain slightly different object representations (Malach et al. 2002). The anterior or ventral part is called *ventral temporo-occipital* area (vTO/VOT, Grill-Spector 2003), but also *posterior fusiform* (pFs, Kourtzi et al. 2003), and responds more invariantly to position and size, suggesting a volumetric 3D object representation (Moore and Engel 2001). The posterior area, the *lateral occipital cortex*, LO, is instead more invariant to the orientation of 2D shapes and to illumination changes. Similar subdivisions have been found in the monkey *inferior temporal* cortex IT (Janssen et al. 2000; Gattass et al. 2005). The *occipito-temporal transition* TEO, which shows a highly invariant response to object identity, is the most likely correspondent of human LOC (Webster et al. 1994; Denys et al. 2004). Although these findings need to be completed and clarified, they suggest a possible mechanisms for object identification and recognition which involves both structural and image-based processing (Tarr and Bülthoff 1998). Object recognition would be achieved integrating, through feature correlation and saliency maps, a partially viewpoint-dependent 3D information (from vTO) with a silhouette classification performed by LO (Kourtzi and Huberle 2005). This solution would solve the long-standing issue on the nature of object recognition, confirming the validity of both the multiple view object representation model (Bülthoff et al. 1995) and the viewpoint invariance hypothesis (James et al. 2002).

Thus, the assumption that ventral stream object representations should be highly viewpoint invariant would not collide with findings suggesting that in active object exploration for recognition subjects search for "preferred" views (James et al. 2001). Similarly, the integrated model would partially explain why 3D orientation response in V4—compatible with both viewpoint dependent and viewpoint invariant models (Hinkle and Connor 2002)—is stimulus-dependent (Hegdé and Essen 2005). On the other hand, object recognition is very likely a gradual process rather than a binary one. Bar et al. (2001) and Grill-Spector et al. (2000) observed that activation in the anterior LOC is modulated by the actual level of recognition, and not by the nature of the stimulus. In any case, geometric data are integrated with additional information, regarding for example color and texture of objects, to speed up and make object recognition more reliable (Humphrey et al. 1994).

Regarding possible connections of ventral stream areas with the intraparietal sulcus, a direct link has been found in the macaque brain between the most 3D responsive ventral inferior temporal area (the lower bank of the superior temporal sulcus) with CIP (Janssen et al. 2000). This link could indicate both a ventral contribution to pose estimation and a dorsal help in object recognition, as explained in Sect. 2.3.1.3.

2.4.2 Other Brain Areas Involved in Grasping

Several areas of the brain not belonging to the two streams are involved in the preparation and/or execution of vision-based grasping actions. A brief description of some very important ones is provided below.

Somatosensory cortex. Located in the anterior part of the parietal cortex, just behind the central sulcus, it is composed of the *primary somatosensory cortex* SI, the equivalent of V1 for the sense of touch, and the secondary somatosensory cortex SII. The former is active in correspondence to any tactile stimulation on the body, the second performs an elaboration of the sensory input in order to detect more complex patterns such as roughness, hardness, compliance estimation in hand haptic exploration (Reed et al. 2004; Newman et al. 2005). Activation of somatosensory areas during grasping is consistently observed (see e.g. Ehrsson et al. 2000; Gardner et al. 2002; Begliomini et al. 2007) and, for what concerns the spatial aspects of grasping and manipulation, higher level processing of tactile information is very likely performed by AIP and nearby areas (Roland et al. 1998).

Prefrontal cortex. The basic role of the prefrontal cortex, PFC, is the organization and orchestration of thoughts and actions in accordance with internal goals and attentional mechanisms (Lebedev et al. 2004). In the specific case of grasping, the PFC is believed to mediate action selection with information on the specific task to perform (Passingham and Toni 2001; Johnson-Frey et al. 2005).

Basal ganglia. Although much is still unrevealed regarding the exact function of this ancient part of the brain, the basal ganglia are probably involved in mediating between rival perceptions and/or competing motor actions. Both area AIP and the ventral premotor cortex receive inputs from the basal ganglia, but from non-overlapping regions, suggesting that they use different selection/evaluation signals that can be used in deciding among candidate target features or hand configurations in grasping (Clower et al. 2005).

Cerebellum. The function of the cerebellum is still under debate. Nevertheless, its involvement in the coordination of action execution and in adaptive sensorimotor control is well recognized (Ramnani et al. 2001; Barlow 2002). In particular, it has been argued that the cerebellum is where internal forward models, largely used in sensorimotor control, are located (Kawato et al. 2003). The cerebellum has an important common feature with the posterior parietal cortex, as both play a role in sensorimotor prediction, more during action execution for the cerebellum, more planning-related for the parietal cortex (Blakemore and Sirigu 2003). Moreover, a considerable component of cerebellar output is devoted to influencing the functional operations of posterior parietal cortex, and neurons of the cerebellum that project to AIP overlap the output channel to the PMv, indicating the existence of a three-way circuit AIP-PMv-Cerebellum (Clower et al. 2005; Tunik et al. 2005).

2.4.3 The Visual Streams in Action

The underlying idea of the original two streams theory (Goodale and Milner 1992; Milner and Goodale 1995) is that visual information has direct control over action in the dorsal stream, without any intervening mental representations. According to this view, neural activity in the dorsal stream does not reflect the representation of objects or events, but rather the direct transformation of visual information into the required coordinates for action. As stated by Jeannerod et al. (1995): "object attributes are processed differently according to the task in which a subject is involved. To serve object-oriented action, these attributes are subjected to a pragmatic mode of processing, the function of which is to extract parameters that are relevant to action, and to generate the corresponding motor commands".

A basic assumption of the processing dualism is that the ventral stream makes use of a contextual coding system for size, distance and orientation of objects, while the dorsal stream needs "real-world" metrics to properly interact with the environment. The patterns of activity of LOC, AIP and other areas of the two streams strongly support this hypothesis, confirming the contrast between the *conceptual* and the *pragmatic* ways of processing of the two streams.

On the other hand, there is a growing experimental evidence for multiple interaction between the streams (Zanon et al. 2010; Verhoef et al. 2011; Perry et al. 2014; Wokke et al. 2014). The original theory has to be constantly updated and suited to new findings (Goodale and Haffenden 2003; Goodale and Westwood 2004; Goodale 2008; Cloutman 2013). For what concerns grasping, there is probably a ventral stream contribution to the grip selection process, through semantic knowledge and memories of past events (Goodale and Milner 2004). Human research demonstrated that choosing a grip depends not only on its visual properties, but also on the meaning we attach to it (Creem and Proffitt 2001; Tankus and Fried 2012) (ventral stream data) and the expected task consequent to the grip (Ansuini et al. 2006; Ikkai et al. 2011) (prefrontal cortex data).

As mentioned in Sect. 2.1.1, visual agnosia patient DF shows normal grasping abilities, as her dorsal stream correctly computes grasping parameters. Nevertheless, her grips on tools are functionally inappropriate, as she does not identify the target object due to her damaged ventral stream, at least until haptic exploration allows her to properly recognize the object (Goodale and Milner 2004). It looks as the decision on exactly "where" to grasp the object can be taken independently from the ventral stream, but with no selection of the object feature or part more suitable for the interaction with the hand, as any semantic meaning of the action is extraneous to the dorsal stream. Moreover, DF cannot scale her grip aperture properly when she has to grasp an object that was removed from view only two seconds earlier (Goodale et al. 1994). According to the authors, this is likely due to the need of accessing object memories stored in, or accessible through, the ventral stream. Patients with optic ataxia exhibit the opposite pattern, as their grasping performances improve if a delay is introduced between target presentation and movement onset, suggesting that memory-mediated action are likely to use different mechanisms in which the dorsal pathway is less critical (Milner et al. 1999, 2001). Psychophysical experiments on delayed grasping in different conditions with normal subjects support the idea that memory-guided grasping relies on the processing of stored information shared with the perception-based ventral system (Singhal et al. 2007).

2.5 The *Third* Stream of Visual Processing

As explained above, AIP is the cortical region in which visual information is used to code an appropriate grasping configuration for a target object, and the detailed parameters of the selected action are determined by processing in the dorsal stream. Nevertheless, action selection is very likely aided by visual processing in the ventral stream. For example, a full object description might be necessary for specifying grip and load forces through estimation or recall of the object weight. Such representation could be used also to avoid grasping objects that can not be grasped, because they are heavy, uncomfortable to handle or even dangerous (Goodale and Humphrey 1998; Westwood et al. 2002).

Considering possible different acting conditions, although some basic grasping movements may be made without the influence of the context or any top-down visual knowledge (Goodale and Milner 2004), in most cases parietal grasp selection is probably driven top-down by semantic information, especially for tools and well known objects (Creem and Proffitt 2001; Frey et al. 2005). In support of this view, Sugio et al. (2003a) showed that different brain areas activate depending on the familiarity with the object, confirming that AIP elaboration is less critical if the object is well known, suggesting that in these cases the ventral stream does most of the job and the action is mainly memory-driven. Other findings (Himmelbach and Karnath 2005) suggest that, although there may be a dramatic shift between the dorsal and ventral systems instantly after the target has disappeared, there also seems to be a progressive change depending on the time delay between target presentation and movement onset.

The above described interaction mechanisms between the streams might be explained by the existence of a so called "third pathway" (Rizzolatti and Matelli 2003; Hoeren et al. 2013). In fact, it has been proposed that the areas of the posterior parietal cortex constituting the classical dorsal pathway should be subdivided into two different sub-systems separated by the intraparietal sulcus (Fogassi and Luppino 2005; Jeannerod and Jacob 2005; Kravitz et al. 2011; Binkofski and Buxbaum 2013). Areas of the *superior parietal lobe* above the IPS, would perform the sensorimotor transformations traditionally assigned to the dorsal stream, related to the online analysis of visual data aimed at generating suitable motor reactions. The second system would contain the *inferior parietal lobe*, below the IPS, including AIP, LIP and regions that seem to be especially human and not matched by structures in other primate's brains (Rushworth et al. 2006). These areas would be dedicated to higher level visuomotor representations, such as those related to the mirror system, which seems to include this part of the PPC (Rizzolatti and Craighero 2004). Indications for a cognitive role of the IPL (and especially AIP) beyond the traditional pragmatic processing have been put forth by several studies (Gallese et al. 1999; Chao and Martin 2000; Derbyshire et al. 2005; Culham and Valyear 2006; Hamilton and Grafton 2006; Durand et al. 2007; Tunik et al. 2007; Schintu et al. 2014). According to this view, a grasp for AIP is a sensorimotor transformation from visual information about the object to motor commands for grasping the object, but also a meaningful action that puts in relation the agent with a feature of the environment. The new, ventro-dorsal stream of the inferior parietal lobe would constitute an ideal convergence focus for the integration of conceptual ventral information with traditional online dorsal data (Rozzi et al. 2006; Gallese 2007). More research is though needed to assess and develop this hypothesis.

This chapter introduced the main neuroscience concepts which will be used and referred to throughout the book. Additional details regarding functions and connectivity of brain areas useful for modeling purposes will be provided in the next chapters.

References

Adams DL, Zeki S (2001) Functional organization of macaque V3 for stereoscopic depth. J Neurophysiol 86(5):2195–2203

Ansuini C, Santello M, Massaccesi S, Castiello U (2006) Effects of end-goal on hand shaping. J Neurophysiol 95(4):2456–2465. doi:10.1152/jn.01107.2005

Ansuini C, Tognin V, Turella L, Castiello U (2007b) Distractor objects affect fingers' angular distances but not fingers' shaping during grasping. Exp Brain Res 178(2):194–205. doi:10.1007/s00221-006-0723-0

Backus BT, Fleet DJ, Parker AJ, Heeger DJ (2001) Human cortical activity correlates with stereoscopic depth perception. J Neurophysiol 86(4):2054–2068

Baizer JS, Ungerleider LG, Desimone R (1991) Organization of visual inputs to the inferior temporal and posterior parietal cortex in macaques. J Neurosci 11(1):168–190

Bar M, Tootell RB, Schacter DL, Greve DN, Fischl B, Mendola JD, Rosen BR, Dale AM (2001) Cortical mechanisms specific to explicit visual object recognition. Neuron 29(2):529–535

Barlow JS (2002) The cerebellum and adaptive control. Cambridge University Press, Cambridge

Begliomini C, Wall MB, Smith AT, Castiello U (2007) Differential cortical activity for precision and whole-hand visually guided grasping in humans. Eur J Neurosci 25(4):1245–1252. doi:10.1111/j.1460-9568.2007.05365.x

Binkofski F, Dohle C, Posse S, Stephan KM, Hefter H, Seitz RJ, Freund HJ (1998) Human anterior intraparietal area subserves prehension: a combined lesion and functional MRI activation study. Neurology 50(5):1253–1259

Binkofski F, Buxbaum LJ (2013) Two action systems in the human brain. Brain Lang 127(2):222–229. doi:10.1016/j.bandl.2012.07.007

Blakemore SJ, Sirigu A (2003) Action prediction in the cerebellum and in the parietal lobe. Exp Brain Res 153(2):239–245. doi:10.1007/s00221-003-1597-z

Borra E, Belmalih A, Calzavara R, Gerbella M, Murata A, Rozzi S, Luppino G (2007) Cortical connections of the macaque anterior intraparietal (AIP) area. Cereb Cortex 18(5):1094–1111

Bremmer F, Schlack A, Duhamel JR, Graf W, Fink GR (2001) Space coding in primate posterior parietal cortex. Neuroimage 14(1 Pt 2):S46–S51. doi:10.1006/nimg.2001.0817

Bremmer F, Klam F, Duhamel JR, Hamed SB, Graf W (2002) Visual-vestibular interactive responses in the macaque ventral intraparietal area (VIP). Eur J Neurosci 16(8):1569–1586

Brouwer GJ, van Ee R, Schwarzbach J (2005) Activation in visual cortex correlates with the awareness of stereoscopic depth. J Neurosci 25(45):10403–10413. doi: 10.1523/JNEUROSCI.2408-05.2005

Bullier J (2001) Integrated model of visual processing. Brain Res Rev 36(2–3):96–107

Bülthoff HH, Edelman SY, Tarr MJ (1995) How are three-dimensional objects represented in the brain? Cereb Cortex 5(3):247–260

Buneo CA, Andersen RA (2006) The posterior parietal cortex: sensorimotor interface for the planning and online control of visually guided movements. Neuropsychologia 44(13):2594–2606. doi:10.1016/j.neuropsychologia.2005.10.011

Buxbaum L, Branch Coslett H (1997) Subtypes of optic ataxia: reframing the disconnection account. Neurocase 3:159–166

Castiello U (2005) The neuroscience of grasping. Nat Rev Neurosci 6(9):726–736. doi:10.1038/nrn1744

Castiello U, Begliomini C (2008) The cortical control of visually guided grasping. Neuroscientist 14(2):157–170. doi:10.1177/1073858407312080

Cavina-Pratesi C, Goodale MA, Culham JC (2007a) fMRI reveals a dissociation between grasping and perceiving the size of real 3D objects. PLoS ONE 2(5):e424. doi:10.1371/journal.pone.0000424

Cavina-Pratesi C, Monaco S, McAdam T, Milner D, Schenk T, Culham JC (2007b) Which aspects of hand-preshaping does human AIP compute during visually guided actions? Evidence from event-related fMRI. In: Annual meeting of the society for neuroscience

Chalupa LM, Werner JS (eds) (2003) The visual neurosciences. MIT Press

Chao LL, Martin A (2000) Representation of manipulable man-made objects in the dorsal stream. Neuroimage 12(4):478–484. doi:10.1006/nimg.2000.0635

Choi HJ, Zilles K, Mohlberg H, Schleicher A, Fink GR, Armstrong E, Amunts K (2006) Cytoarchitectonic identification and probabilistic mapping of two distinct areas within the anterior ventral bank of the human intraparietal sulcus. J Comp Neurol 495(1):53–69. doi:10.1002/cne.20849

Christel MI, Billard A (2002) Comparison between macaques' and humans' kinematics of prehension: the role of morphological differences and control mechanisms. Behav Brain Res 131(1–2):169–184

Cloutman LL (2013) Interaction between dorsal and ventral processing streams: where, when and how? Brain Lang 127(2):251–263. doi:10.1016/j.bandl.2012.08.003

Clower DM, Dum RP, Strick PL (2005) Basal ganglia and cerebellar inputs to 'AIP'. Cereb Cortex 15(7):913–920. doi:10.1093/cercor/bhh190

Cohen NR, Cross ES, Tunik E, Grafton ST, Culham JC (2009) Ventral and dorsal stream contributions to the online control of immediate and delayed grasping: a tms approach. Neuropsychologia 47(6):1553–1562. doi:10.1016/j.neuropsychologia.2008.12.034

Creem SH, Proffitt DR (2001) Grasping objects by their handles: a necessary interaction between cognition and action. J Exp Psychol Hum Percept Perform 27(1):218–228

Culham JC (2001) How neurons become BOLD? Trends Cogn Sci 5(10):416

Culham JC, Danckert SL, DeSouza JFX, Gati JS, Menon RS, Goodale MA (2003) Visually guided grasping produces fMRI activation in dorsal but not ventral stream brain areas. Exp Brain Res 153(2):180–189. doi:10.1007/s00221-003-1591-5

Culham JC (2004) Human brain imaging reveals a parietal area specialized for grasping. In: Kanwisher N, Duncan J (eds) Functional neuroimaging of visual cognition: attention and performance XX. Oxford University Press, Oxford, pp 417–438

Culham JC, Cavina-Pratesi C, Singhal A (2006) The role of parietal cortex in visuomotor control: what have we learned from neuroimaging? Neuropsychologia 44(13):2668–2684. doi:10.1016/j.neuropsychologia.2005.11.003

Culham JC, Valyear KF (2006) Human parietal cortex in action. Curr Opin Neurobiol 16(2):205–212. doi:10.1016/j.conb.2006.03.005

Cumming BG, DeAngelis GC (2001) The physiology of stereopsis. Annu Rev Neurosci 24:203–238. doi:10.1146/annurev.neuro.24.1.203

Darling S, Della Sala S, Logie RH, Cantagallo A (2006) Neuropsychological evidence for separating components of visuospatial working memory. J Neurol 253(2):176–180. doi:10.1007/s00415-005-0944-3

Davare M, Andres M, Cosnard G, Thonnard JL, Olivier E (2006) Dissociating the role of ventral and dorsal premotor cortex in precision grasping. J Neurosci 26(8):2260–2268. doi:10.1523/JNEUROSCI.3386-05.2006

Davare M, Zénon A, Pourtois G, Desmurget M, Olivier E (2012) Role of the medial part of the intraparietal sulcus in implementing movement direction. Cereb Cortex 22(6):1382–1394. doi:10.1093/cercor/bhr210

de Vries SC, Kappers AM, Koenderink JJ (1994) Influence of surface attitude and curvature scaling on discrimination of binocularly presented curved surfaces. Vision Res 34(18):2409–2423

DeAngelis GC, Cumming BG, Newsome WT (1998) Cortical area MT and the perception of stereoscopic depth. Nature 394(6694):677–680. doi:10.1038/29299

Debowy DJ, Ghosh S, Ro JY, Gardner EP (2001) Comparison of neuronal firing rates in somatosensory and posterior parietal cortex during prehension. Exp Brain Res 137(3–4):269–291

Denys K, Vanduffel W, Fize D, Nelissen K, Peuskens H, Essen DV, Orban GA (2004) The processing of visual shape in the cerebral cortex of human and nonhuman primates: a functional magnetic resonance imaging study. J Neurosci 24(10):2551–2565. doi:10.1523/JNEUROSCI.3569-03.2004

Derbyshire N, Ellis R, Tucker M (2005) The potentiation of two components of the reach-to-grasp action during object categorisation in visual memory. ACTA Psychol (Amsterdam) 122:74–98. doi:10.1016/j.actpsy.2005.10.004

Di Pellegrino G, Fadiga L, Fogassi L, Gallese V, Rizzolatti G (1992) Understanding motor events: a neurophysiological study. Exp Brain Res 91(1):176–180

Durand JB, Nelissen K, Joly O, Wardak C, Todd JT, Norman JF, Janssen P, Vanduffel W, Orban GA (2007) Anterior regions of monkey parietal cortex process visual 3D shape. Neuron 55(3):493–505. doi:10.1016/j.neuron.2007.06.040

Ehrsson HH, Fagergren A, Jonsson T, Westling G, Johansson RS, Forssberg H (2000) Cortical activity in precision- versus power-grip tasks: an fMRI study. J Neurophysiol 83(1):528–536

Ellison A, Cowey A (2009) Differential and co-involvement of areas of the temporal and parietal streams in visual tasks. Neuropsychologia 47(6):1609–1614. doi:10.1016/j.neuropsychologia.2008.12.013

Farah MJ (2004) Visual agnosia. MIT Press
Felleman DJ, Essen DCV (1991) Distributed hierarchical processing in the primate cerebral cortex. Cereb Cortex 1(1):1–47
Ferrera VP, Nealey TA, Maunsell JH (1992) Mixed parvocellular and magnocellular geniculate signals in visual area V4. Nature 358(6389):756–761. doi:10.1038/358756a0
Fogassi L, Ferrari PF, Gesierich B, Rozzi S, Chersi F, Rizzolatti G (2005) Parietal lobe: from action organization to intention understanding. Science 308(5722):662–667. doi:10.1126/science.1106138
Fogassi L, Luppino G (2005) Motor functions of the parietal lobe. Curr Opin Neurobiol 15(6):626–631. doi:10.1016/j.conb.2005.10.015
Frey SH, Vinton D, Norlund R, Grafton ST (2005) Cortical topography of human anterior intraparietal cortex active during visually guided grasping. Cogn Brain Res 23(2–3):397–405. doi:10.1016/j.cogbrainres.2004.11.010
Gallese V (2007) The "conscious" dorsal stream: embodied simulation and its role in space and action conscious awareness. Psychedelic 13(1):1–20
Gallese V, Craighero L, Fadiga L, Fogassi L (1999) Perception through action. Psyche 5(21):1
Ganel T, Goodale MA (2003) Visual control of action but not perception requires analytical processing of object shape. Nature 426(6967):664–667. doi:10.1038/nature02156
Gardner EP, Debowy DJ, Ro JY, Ghosh S, Babu KS (2002) Sensory monitoring of prehension in the parietal lobe: a study using digital video. Behav Brain Res 135(1–2):213–224
Gattass R, Nascimento-Silva S, Soares JGM, Lima B, Jansen AK, Diogo ACM, Farias MF, Botelho MMEP, Mariani OS, Azzi J, Fiorani M (2005) Cortical visual areas in monkeys: location, topography, connections, columns, plasticity and cortical dynamics. Philos Trans R Soc B: Biol Sci 360(1456):709–731. doi:10.1098/rstb.2005.1629
Gegenfurtner K, Kiper D, Levitt J (1997) Functional properties of neurons in macaque area V3. J Neurophysiol 77(4):1906–1923
Genovesio A, Ferraina S (2004) Integration of retinal disparity and fixation-distance related signals toward an egocentric coding of distance in the posterior parietal cortex of primates. J Neurophysiol 91(6):2670–2684. doi:10.1152/jn.00712.2003
Gibson JJ (1979) The ecological approach to visual perception. Lawrence Erlbaum Associates, New Jersey, USA
Glover S (2003) Optic ataxia as a deficit specific to the on-line control of actions. Neurosci Biobehav Rev 27(5):447–456
Glover S, Miall RC, Rushworth MFS (2005) Parietal rTMS disrupts the initiation but not the execution of on-line adjustments to a perturbation of object size. J Cogn Neurosci 17(1):124–136. doi:10.1162/0898929052880066
Goodale MA, Milner AD (2004) Sight unseen. Oxford University Press, Oxford
Goodale MA, Milner AD, Jakobson LS, Carey DP (1991) A neurological dissociation between perceiving objects and grasping them. Nature 349(6305):154–156. doi:10.1038/349154a0
Goodale MA, Meenan JP, Bülthoff HH, Nicolle DA, Murphy KJ, Racicot CI (1994) Separate neural pathways for the visual analysis of object shape in perception and prehension. Curr Biol 4(7):604–610
Goodale MA (2008) Action without perception in human vision. Cogn Neuropsychol 25(7–8):891–919. doi:10.1080/02643290801961984
Goodale MA, Haffenden AM (2003) Interactions between the dorsal and ventral streams of visual processing. Adv Neurol 93:249–267
Goodale MA, Humphrey GK (1998) The objects of action and perception. Cognition 67(1–2):181–207
Goodale MA, Milner AD (1992) Separate visual pathways for perception and action. Trends Neurosci 15(1):20–25
Goodale MA, Westwood DA (2004) An evolving view of duplex vision: separate but interacting cortical pathways for perception and action. Curr Opin Neurobiol 14(2):203–211. doi:10.1016/j.conb.2004.03.002

Gréa H, Pisella L, Rossetti Y, Desmurget M, Tilikete C, Grafton S, Prablanc C, Vighetto A (2002) A lesion of the posterior parietal cortex disrupts on-line adjustments during aiming movements. Neuropsychologia 40(13):2471–2480

Grefkes C, Weiss PH, Zilles K, Fink GR (2002) Crossmodal processing of object features in human anterior intraparietal cortex: an fMRI study implies equivalencies between humans and monkeys. Neuron 35(1):173–184

Grefkes C, Ritzl A, Zilles K, Fink GR (2004) Human medial intraparietal cortex subserves visuomotor coordinate transformation. Neuroimage 23(4):1494–1506. doi:10.1016/j.neuroimage.2004.08.031

Grefkes C, Fink GR (2005) The functional organization of the intraparietal sulcus in humans and monkeys. J Anat 207(1):3–17. doi:10.1111/j.1469-7580.2005.00426.x

Grill-Spector K, Kushnir T, Hendler T, Edelman S, Itzchak Y, Malach R (1998) A sequence of object-processing stages revealed by fMRI in the human occipital lobe. Hum Brain Mapp 6(4):316–328

Grill-Spector K, Kushnir T, Edelman S, Avidan G, Itzchak Y, Malach R (1999) Differential processing of objects under various viewing conditions in the human lateral occipital complex. Neuron 24(1):187–203

Grill-Spector K, Kushnir T, Hendler T, Malach R (2000) The dynamics of object-selective activation correlate with recognition performance in humans. Nat Neurosci 3(8):837–843. doi:10.1038/77754

Grill-Spector K, Kourtzi Z, Kanwisher N (2001) The lateral occipital complex and its role in object recognition. Vision Res 41(10–11):1409–1422

Grill-Spector K (2003) The neural basis of object perception. Curr Opin Neurobiol 13(2):159–166

Hamilton AF, Grafton ST (2006) Goal representation in human anterior intraparietal sulcus. J Neurosci 26(4):1133–1137. doi:10.1523/JNEUROSCI.4551-05.2006

Heeger DJ, Ress D (2002) What does fMRI tell us about neuronal activity? Nat Rev Neurosci 3(2):142–151. doi:10.1038/nrn730

Hegdé J, Essen DCV (2005) Stimulus dependence of disparity coding in primate visual area V4. J Neurophysiol 93(1):620–626. doi:10.1152/jn.00039.2004

Himmelbach M, Karnath HO (2005) Dorsal and ventral stream interaction: contributions from optic ataxia. J Cogn Neurosci 17(4):632–640. doi:10.1162/0898929053467514

Hinkle DA, Connor CE (2002) Three-dimensional orientation tuning in macaque area V4. Nat Neurosci 5(7):665–670

Hoeren M, Kaller CP, Glauche V, Vry MS, Rijntjes M, Hamzei F, Weiller C (2013) Action semantics and movement characteristics engage distinct processing streams during the observation of tool use. Exp Brain Res 229(2):243–260. doi:10.1007/s00221-013-3610-5

Humphrey GK, Goodale MA, Jakobson LS, Servos P (1994) The role of surface information in object recognition: studies of a visual form agnosic and normal subjects. Perception 23(12):1457–1481

Ikkai A, Jerde TA, Curtis CE (2011) Perception and action selection dissociate human ventral and dorsal cortex. J Cogn Neurosci 23(6):1494–1506. doi:10.1162/jocn.2010.21499

James KH, Humphrey GK, Goodale MA (2001) Manipulating and recognizing virtual objects: where the action is. Can J Exp Psychol 55(2):111–120

James T, Humphrey G, Gati J, Menon R, Goodale M (2002) Differential effects of viewpoint on object-driven activation in dorsal and ventral streams. Neuron 35(4):793–801

James TW, Culham J, Humphrey GK, Milner AD, Goodale MA (2003) Ventral occipital lesions impair object recognition but not object-directed grasping: an fMRI study. Brain 126(Pt 11):2463–2475. doi:10.1093/brain/awg248

Jäncke L, Kleinschmidt A, Mirzazade S, Shah NJ, Freund HJ (2001) The role of the inferior parietal cortex in linking the tactile perception and manual construction of object shapes. Cereb Cortex 11(2):114–121

Janssen P, Vogels R, Orban GA (2000) Selectivity for 3D shape that reveals distinct areas within macaque inferior temporal cortex. Science 288(5473):2054–2056

Jeannerod M (1997) The cognitive neuroscience of action. Blackwell, Oxford

Jeannerod M, Arbib MA, Rizzolatti G, Sakata H (1995) Grasping objects: the cortical mechanisms of visuomotor transformation. Trends Neurosci 18(7):314–320
Jeannerod M, Jacob P (2005) Visual cognition: a new look at the two-visual systems model. Neuropsychologia 43(2):301–312. doi:10.1016/j.neuropsychologia.2004.11.016
Johnson-Frey SH, Newman-Norlund R, Grafton ST (2005) A distributed left hemisphere network active during planning of everyday tool use skills. Cereb Cortex 15(6):681–695. doi:10.1093/cercor/bhh169
Kalaska JF, Cisek P, Gosselin-Kessiby N (2003) Mechanisms of selection and guidance of reaching movements in the parietal lobe. Adv Neurol 93:97–119
Katsuyama N, Naganuma T, Sakata H, Taira M (2005) Coding of 3D curvature in the parietal cortex (area CIP) of macaque monkey. In: International symposium on autonomous minirobots for research and edutainment, pp 181–186
Kawasaki M, Watanabe M, Okuda J, Sakagami M, Aihara K (2008) Human posterior parietal cortex maintains color, shape and motion in visual short-term memory. Brain Res 1213:91–97. doi:10.1016/j.brainres.2008.03.037
Kawato M, Kuroda T, Imamizu H, Nakano E, Miyauchi S, Yoshioka T (2003) Internal forward models in the cerebellum: fMRI study on grip force and load force coupling. Prog Brain Res 142:171–188
Kourtzi Z, Erb M, Grodd W, Bülthoff HH (2003) Representation of the perceived 3-D object shape in the human lateral occipital complex. Cereb Cortex 13(9):911–920
Kourtzi Z, Huberle E (2005) Spatiotemporal characteristics of form analysis in the human visual cortex revealed by rapid event-related fMRI adaptation. Neuroimage 28(2):440–452. doi:10.1016/j.neuroimage.2005.06.017
Kourtzi Z, Kanwisher N (2000) Cortical regions involved in perceiving object shape. J Neurosci 20(9):3310–3318
Kourtzi Z, Kanwisher N (2001) Representation of perceived object shape by the human lateral occipital complex. Science 293(5534):1506–1509. doi:10.1126/science.1061133
Kravitz DJ, Saleem KS, Baker CI, Mishkin M (2011) A new neural framework for visuospatial processing. Nat Rev Neurosci 12(4):217–230. doi:10.1038/nrn3008
Lebedev MA, Messinger A, Kralik JD, Wise SP (2004) Representation of attended versus remembered locations in prefrontal cortex. PLoS Biol 2(11):e365. doi:10.1371/journal.pbio.0020365
Lee TS (2003) Computations in the early visual cortex. J Physiol Paris 97(2–3):121–139. doi:10.1016/j.jphysparis.2003.09.015
Lewis JW, Essen DCV (2000) Corticocortical connections of visual, sensorimotor, and multimodal processing areas in the parietal lobe of the macaque monkey. J Comp Neurol 428(1):112–137
Logothetis NK (1999) Vision: a window on consciousness. Sci Am 281(5):69–75
Logothetis NK, Pauls J, Augath M, Trinath T, Oeltermann A (2001) Neurophysiological investigation of the basis of the fMRI signal. Nature 412(6843):150–157. doi:10.1038/35084005
Luppino G, Murata A, Govoni P, Matelli M (1999) Largely segregated parietofrontal connections linking rostral intraparietal cortex (areas AIP and VIP) and the ventral premotor cortex (areas F5 and F4). Exp Brain Res 128(1–2):181–187
Luppino G, Rizzolatti G (2000) The organization of the frontal motor cortex. News Physiol Sci 15:219–224
Makuuchi M, Someya Y, Ogawa S, Takayama Y (2012) Hand shape selection in pantomimed grasping: interaction between the dorsal and the ventral visual streams and convergence on the ventral premotor area. Hum Brain Mapp 33(8):1821–1833. doi:10.1002/hbm.21323
Malach R, Levy I, Hasson U (2002) The topography of high-order human object areas. Trends Cogn Sci 6(4):176–184
Maunsell JH, (1992) Functional visual streams. Curr Opin Neurobiol 2(4):506–510
Milner AD, Goodale MA (1995) The visual brain in action. Oxford University Press, Oxford
Milner AD, Perrett DI, Johnston RS, Benson PJ, Jordan TR, Heeley DW, Bettucci D, Mortara F, Mutani R, Terazzi E (1991) Perception and action in 'visual form agnosia'. Brain 114(Pt 1B):405–428

Milner AD, Paulignan Y, Dijkerman HC, Michel F, Jeannerod M (1999) A paradoxical improvement of misreaching in optic ataxia: new evidence for two separate neural systems for visual localization. Proc R Soc B: Biol Sci 266(1434):2225–2229
Milner AD, Dijkerman HC, Pisella L, McIntosh RD, Tilikete C, Vighetto A, Rossetti Y (2001) Grasping the past. Delay can improve visuomotor performance. Curr Biol 11(23):1896–1901
Milner AD, Goodale MA (1993) Visual pathways to perception and action. Prog Brain Res 95:317–337
Moore C, Engel S (2001) Neural response to the perception of volume in the lateral occipital complex. Neuron 29(1):277–286
Murata A, Fadiga L, Fogassi L, Gallese V, Raos V, Rizzolatti G (1997) Object representation in the ventral premotor cortex (area F5) of the monkey. J Neurophysiol 78(4):2226–2230
Murata A, Gallese V, Luppino G, Kaseda M, Sakata H (2000) Selectivity for the shape, size, and orientation of objects for grasping in neurons of monkey parietal area AIP. J Neurophysiol 83(5):2580–2601
Naganuma T, Nose I, Inoue K, Takemoto A, Katsuyama N, Taira M (2005) Information processing of geometrical features of a surface based on binocular disparity cues: an fMRI study. Neurosci Res 51(2):147–155. doi:10.1016/j.neures.2004.10.009
Newman SD, Klatzky RL, Lederman SJ, Just MA (2005) Imagining material versus geometric properties of objects: an fMRI study. Cogn Brain Res 23(2–3):235–246. doi:10.1016/j.cogbrainres.2004.10.020
Nguyenkim JD, DeAngelis GC (2003) Disparity-based coding of three-dimensional surface orientation by macaque middle temporal neurons. J Neurosci 23(18):7117–7128
Orban GA, Fize D, Peuskens H, Denys K, Nelissen K, Sunaert S, Todd J, Vanduffel W (2003) Similarities and differences in motion processing between the human and macaque brain: evidence from fMRI. Neuropsychologia 41(13):1757–1768
Orban GA, Claeys K, Nelissen K, Smans R, Sunaert S, Todd JT, Wardak C, Durand JB, Vanduffel W (2006a) Mapping the parietal cortex of human and non-human primates. Neuropsychologia 44(13):2647–2667. doi:10.1016/j.neuropsychologia.2005.11.001
Orban GA, Janssen P, Vogels R (2006) Extracting 3D structure from disparity. Trends Neurosci 29(8):466–473. doi:10.1016/j.tins.2006.06.012
Orban GA, Caruana F (2014) The neural basis of human tool use. Front Psychol 5:310. doi:10.3389/fpsyg.2014.00310
Parker AJ (2004) From binocular disparity to the perception of stereoscopic depth. In: Chalupa LM, Werner JS (eds) The visual neurosciences, Chap 49. MIT Press, Cambridge, MA, pp 779–792
Pascual-Leone A, Davey NJ, Rothwell J, Wasserman E, Puri BK (eds) (2002) Handbook of transcranial magnetic stimulation. Arnold
Passingham RE, Toni I (2001) Contrasting the dorsal and ventral visual systems: guidance of movement versus decision making. Neuroimage 14(1 Pt 2):S125–S131. doi:10.1006/nimg.2001.0836
Perry CJ, Tahiri A, Fallah M (2014) Feature integration within and across visual streams occurs at different visual processing stages. J Vis 14(2): doi:10.1167/14.2.10
Pitzalis S, Sereno MI, Committeri G, Fattori P, Galati G, Tosoni A, Galletti C (2013) The human homologue of macaque area v6a. Neuroimage 82:517–530. doi:10.1016/j.neuroimage.2013.06.026
Plewan T, Weidner R, Eickhoff SB, Fink GR (2012) Ventral and dorsal stream interactions during the perception of the müller-lyer illusion: evidence derived from fmri and dynamic causal modeling. J Cogn Neurosci 24(10):2015–2029
Poggio GF, Gonzalez F, Krause F (1988) Stereoscopic mechanisms in monkey visual cortex: binocular correlation and disparity selectivity. J Neurosci 8(12):4531–4550
Ramnani N, Toni I, Passingham RE, Haggard P (2001) The cerebellum and parietal cortex play a specific role in coordination: a PET study. Neuroimage 14(4):899–911. doi:10.1006/nimg.2001.0885

Raos V, Umiltà MA, Murata A, Fogassi L, Gallese V (2006) Functional properties of grasping-related neurons in the ventral premotor area F5 of the macaque monkey. J Neurophysiol 95(2):709–729. doi:10.1152/jn.00463.2005
Red SD, Patel SS, Sereno AB (2012) Shape effects on reflexive spatial attention are driven by the dorsal stream. Vision Res 55:32–40. doi:10.1016/j.visres.2011.12.007
Reed CL, Shoham S, Halgren E (2004) Neural substrates of tactile object recognition: an fMRI study. Hum Brain Mapp 21(4):236–246. doi:10.1002/hbm.10162
Reichenbach A, Thielscher A, Peer A, Bülthoff HH, Bresciani JP (2014) A key region in the human parietal cortex for processing proprioceptive hand feedback during reaching movements. Neuroimage 84:615–625. doi:10.1016/j.neuroimage.2013.09.024
Rice NJ, Tunik E, Grafton ST (2006) The anterior intraparietal sulcus mediates grasp execution, independent of requirement to update: new insights from transcranial magnetic stimulation. J Neurosci 26(31):8176–8182. doi:10.1523/JNEUROSCI.1641-06.2006
Rice NJ, McIntosh RD, Schindler I, Mon-Williams M, Dmonet JF, Milner AD (2006a) Intact automatic avoidance of obstacles in patients with visual form agnosia. Exp Brain Res 174(1):176–188. doi:10.1007/s00221-006-0435-5
Rizzolatti G, Camarda R, Fogassi L, Gentilucci M, Luppino G, Matelli M (1988) Functional organization of inferior area 6 in the macaque monkey II. Area F5 and the control of distal movements. Exp Brain Res 71(3):491–507
Rizzolatti G, Fadiga L, Gallese V, Fogassi L (1996) Premotor cortex and the recognition of motor actions. Cogn Brain Res 3(2):131–141
Rizzolatti G, Luppino G, Matelli M (1998) The organization of the cortical motor system: new concepts. Electroencephalogr Clin Neurophysiol 106(4):283–296
Rizzolatti G, Arbib MA (1998) Language within our grasp. Trends Neurosci 21:188–194
Rizzolatti G, Craighero L (2004) The mirror-neuron system. Annu Rev Neurosci 27:169–192. doi:10.1146/annurev.neuro.27.070203.144230
Rizzolatti G, Luppino G (2001) The cortical motor system. Neuron 31(6):889–901
Rizzolatti G, Matelli M (2003) Two different streams form the dorsal visual system: anatomy and functions. Exp Brain Res 153(2):146–157. doi:10.1007/s00221-003-1588-0
Ro JY, Debowy D, Ghosh S, Gardner EP (2000) Depression of neuronal firing rates in somatosensory and posterior parietal cortex during object acquisition in a prehension task. Exp Brain Res 135(1):1–11
Roland PE, O'Sullivan B, Kawashima R (1998) Shape and roughness activate different somatosensory areas in the human brain. Proc Nat Acad Sci USA 95(6):3295–3300
Rozzi S, Calzavara R, Belmalih A, Borra E, Gregoriou GG, Matelli M, Luppino G (2006) Cortical connections of the inferior parietal cortical convexity of the macaque monkey. Cereb Cortex 16(10):1389–1417. doi:10.1093/cercor/bhj076
Rushworth MFS, Behrens TEJ, Johansen-Berg H (2006) Connection patterns distinguish 3 regions of human parietal cortex. Cereb Cortex 16(10):1418–1430. doi:10.1093/cercor/bhj079
Rutschmann RM, Greenlee MW (2004) Bold response in dorsal areas varies with relative disparity level. Neuroreport 15(4):615–619
Sakata H, Taira M, Murata A, Mine S (1995) Neural mechanisms of visual guidance of hand action in the parietal cortex of the monkey. Cereb Cortex 5(5):429–438
Sakata H, Taira M, Kusunoki M, Murata A, Tanaka Y (1997) The TINS lecture. The parietal association cortex in depth perception and visual control of hand action. Trends Neurosci 20(8):350–357
Sakata H, Taira M, Kusunoki M, Murata A, Tanaka Y, Tsutsui K (1998) Neural coding of 3D features of objects for hand action in the parietal cortex of the monkey. Philos Trans R Soc B: Biol Sci 353(1373):1363–1373
Sakata H, Taira M, Kusunoki M, Murata A, Tsutsui K, Tanaka Y, Shein WN, Miyashita Y (1999) Neural representation of three-dimensional features of manipulation objects with stereopsis. Exp Brain Res 128(1–2):160–169

Sakata H, Tsutsui KI, Taira M (2005) Toward an understanding of the neural processing for 3D shape perception. Neuropsychologia 43(2):151–161. doi:10.1016/j.neuropsychologia.2004.11.003

Schenk T, Ellison A, Rice N, Milner AD (2005) The role of V5/MT+ in the control of catching movements: an rTMS study. Neuropsychologia 43(2):189–198. doi:10.1016/j.neuropsychologia.2004.11.006

Scherberger H, Anderson R (2004) Sensorimotor transformation in the posterior parietal cortex. In: Chalupa LM, Werner JS (eds) The visual neurosciences. MIT Press, Cambridge

Schintu S, Hadj-Bouziane F, Dal Monte O, Knutson KM, Pardini M, Wassermann EM, Grafman J, Krueger F (2014) Object and space perception - is it a matter of hemisphere? Cortex 57:244–253. doi:10.1016/j.cortex.2014.04.009

Sereno AB, Sereno ME, Lehky SR (2014) Recovering stimulus locations using populations of eye-position modulated neurons in dorsal and ventral visual streams of non-human primates. Front Integr Neurosci 8:28. doi:10.3389/fnint.2014.00028

Shikata E, Tanaka Y, Nakamura H, Taira M, Sakata H (1996) Selectivity of the parietal visual neurones in 3D orientation of surface of stereoscopic stimuli. Neuroreport 7(14):2389–2394

Shikata E, Hamzei F, Glauche V, Knab R, Dettmers C, Weiller C, Büchel C (2001) Surface orientation discrimination activates caudal and anterior intraparietal sulcus in humans: an event-related fMRI study. J Neurophysiol 85(3):1309–1314

Shikata E, Hamzei F, Glauche V, Koch M, Weiller C, Binkofski F, Büchel C (2003) Functional properties and interaction of the anterior and posterior intraparietal areas in humans. Eur J Neurosci 17(5):1105–1110

Shmuelof L, Zohary E (2005) Dissociation between ventral and dorsal fMRI activation during object and action recognition. Neuron 47(3):457–470. doi:10.1016/j.neuron.2005.06.034

Singhal A, Culham JC, Chinellato E, Goodale MA (2007) Dual-task interference is greater in delayed grasping than in visually guided grasping. J Vis 7(5):1–12

Snow JC, Pettypiece CE, McAdam TD, McLean AD, Stroman PW, Goodale MA, Culham JC (2011) Bringing the real world into the fmri scanner: repetition effects for pictures versus real objects. Sci Rep 1:130. doi:10.1038/srep00130

Striem-Amit E, Dakwar O, Reich L, Amedi A (2012) The large-scale organization of "visual" streams emerges without visual experience. Cereb Cortex 22(7):1698–1709. doi:10.1093/cercor/bhr253

Sugio T, Inui T, Matsuo K, Matsuzawa M, Glover GH, Nakai T (1999) The role of the posterior parietal cortex in human object recognition: a functional magnetic resonance imaging study. Neurosci Lett 276(1):45–48

Sugio T, Ogawa K, Inui T (2003a) Multiple action representations of familiar objects with handles: An fMRI study. In: European conference on visual perception

Taira M, Mine S, Georgopoulos AP, Murata A, Sakata H (1990) Parietal cortex neurons of the monkey related to the visual guidance of hand movement. Exp Brain Res 83(1):29–36

Taira M, Tsutsui KI, Jiang M, Yara K, Sakata H (2000) Parietal neurons represent surface orientation from the gradient of binocular disparity. J Neurophysiol 83(5):3140–3146

Tankus A, Fried I (2012) Visuomotor coordination and motor representation by human temporal lobe neurons. J Cogn Neurosci 24(3):600–610

Tarr MJ, Bülthoff HH (1998) Image-based object recognition in man, monkey and machine. Cognition 67(1–2):1–20

Thoma V, Henson RN (2011) Object representations in ventral and dorsal visual streams: fmri repetition effects depend on attention and part-whole configuration. Neuroimage 57(2):513–525. doi:10.1016/j.neuroimage.2011.04.035

Todd JT (2004) The visual perception of 3D shape. Trends in Cogn Sci 8(3):115–121. doi:10.1016/j.tics.2004.01.006

Tootell RBH, Tsao D, Vanduffel W (2003) Neuroimaging weighs. In: humans meet macaques in "primate" visual cortex. J Neurosci 23(10):3981–3989

Tootell RBH, Mendola JD, Hadjikhani NK, Ledden PJ, Liu AK, Reppas JB, Sereno MI, Dale AM (1997) Functional analysis of V3A and related areas in human visual cortex. J Neurosci 17(18):7060–7078

Tsao DY, Vanduffel W, Sasaki Y, Fize D, Knutsen TA, Mandeville JB, Wald LL, Dale AM, Rosen BR, Essen DCV, Livingstone MS, Orban GA, Tootell RBH (2003) Stereopsis activates V3A and caudal intraparietal areas in macaques and humans. Neuron 39(3):555–568

Tsutsui K, Jiang M, Yara K, Sakata H, Taira M (2001) Integration of perspective and disparity cues in surface-orientation-selective neurons of area CIP. J Neurophysiol 86(6):2856–2867

Tsutsui KI, Jiang M, Sakata H, Taira M (2003) Short-term memory and perceptual decision for three-dimensional visual features in the caudal intraparietal sulcus (area CIP). J Neurosci 23(13):5486–5495

Tsutsui KI, Taira M, Sakata H (2005) Neural mechanisms of three-dimensional vision. Neurosci Res 51(3):221–229. doi:10.1016/j.neures.2004.11.006

Tunik E, Frey SH, Grafton ST (2005) Virtual lesions of the anterior intraparietal area disrupt goal-dependent on-line adjustments of grasp. Nat Neurosci 8(4):505–511. doi:10.1038/nn1430

Tunik E, Rice NJ, Hamilton A, Grafton ST (2007) Beyond grasping: representation of action in human anterior intraparietal sulcus. Neuroimage 36(Suppl 2):T77–T86. doi:10.1016/j.neuroimage.2007.03.026

Ungerleider L, Mishkin M (1982) Two cortical visual systems. In: Goodale M, Mansfield R (eds) Ingle D. Analysis of visual behavior, MIT Press, pp 549–586

Valyear KF, Culham JC, Sharif N, Westwood D, Goodale MA (2006) A double dissociation between sensitivity to changes in object identity and object orientation in the ventral and dorsal visual streams: a human fMRI study. Neuropsychologia 44(2):218–228. doi:10.1016/j.neuropsychologia.2005.05.004

Van Essen DC, Lewis JW, Drury HA, Hadjikhani N, Tootell RB, Bakircioglu M, Miller MI (2001) Mapping visual cortex in monkeys and humans using surface-based atlases. Vision Res 41(10–11):1359–1378

Vanduffel W, Fize D, Peuskens H, Denys K, Sunaert S, Todd JT, Orban GA (2002) Extracting 3D from motion: differences in human and monkey intraparietal cortex. Science 298(5592):413–415. doi:10.1126/science.1073574

Vanni S, Dojat M, Warnking J, Delon-Martin C, Segebarth C, Bullier J (2004) Timing of interactions across the visual field in the human cortex. Neuroimage 21(3):818–828. doi:10.1016/j.neuroimage.2003.10.035

Verhoef BE, Vogels R, Janssen P (2011) Synchronization between the end stages of the dorsal and the ventral visual stream. J Neurophysiol 105(5):2030–2042. doi:10.1152/jn.00924.2010

Webster MJ, Bachevalier J, Ungerleider LG (1994) Connections of inferior temporal areas TEO and TE with parietal and frontal cortex in macaque monkeys. Cereb Cortex 4(5):470–483

Welchman AE, Deubelius A, Conrad V, Bülthoff HH, Kourtzi Z (2005) 3D shape perception from combined depth cues in human visual cortex. Nat Neurosci 8(6):820–827. doi:10.1038/nn1461

Westwood DA, Danckert J, Servos P, Goodale MA (2002) Grasping two-dimensional images and three-dimensional objects in visual-form agnosia. Exp Brain Res 144(2):262–267. doi:10.1007/s00221-002-1068-y

Westwood DA, Goodale MA (2003a) A haptic size-contrast illusion affects size perception but not grasping. Exp Brain Res 153(2):253–259. doi:10.1007/s00221-003-1599-x

Whitwell RL, Milner AD, Goodale MA (2014) The two visual systems hypothesis: new challenges and insights from visual form agnosic patient df. Front Neurol 5:255. doi:10.3389/fneur.2014.00255

Williams J, Whiten A, Suddendorf T, Perrett D (2001) Imitation, mirror neurons and autism. Neurosci Biobehav Rev 25:287–295

Winkler A, Wright CE, Chubb C (2005) Dissociating the functions of visual pathways using equisalient stimuli. J Vis 5(8):362. doi:10.1167/5.8.362

Wokke ME, Scholte HS, Lamme VAF (2014) Opposing dorsal/ventral stream dynamics during figure-ground segregation. J Cogn Neurosci 26(2):365–379

Zachariou V, Klatzky R, Behrmann M (2014) Ventral and dorsal visual stream contributions to the perception of object shape and object location. J Cogn Neurosci 26(1):189–209

Zanon M, Busan P, Monti F, Pizzolato G, Battaglini PP (2010) Cortical connections between dorsal and ventral visual streams in humans: Evidence by tms/eeg co-registration. Brain Topogr 22(4):307–317. doi:10.1007/s10548-009-0103-8

Chapter 3
Intelligent Robotic Grasping?

Mutual interest between the fields of robotics and cognitive sciences has been steadily growing in the recent years, especially through the bridging of artificial intelligence research. Nevertheless, the differences in goals and methodology and the lack of a common language make of true interdisciplinary research still a pioneering work. As the following review will expose, grasping is not an exception to this situation.

A brief description of traditional and bio-inspired research in robotic vision-based grasping is presented and critically discussed in this chapter, with the purpose of defining a few important guidelines required to achieve fruitful cross-disciplinary research. The chapter includes a proposal for grounding symbolic representations in sensorimotor interactions for robotic grasping, synthesized in Sect. 3.3.

3.1 Vision-Based Robotic Grasping, A Brief Outline

The field of robot grasping and manipulation has been steadily growing and developing, as can be noticed comparing the most important research of the near past. The fundamental studies of the eighties (Cutkosky 1985; Mason and Salisbury 1985; Nguyen 1988) defined the basic concepts, both physical and technological, on which most of the later research built upon. In parallel, classical works on grasping in humans and primates were published (Napier 1983; Iberall 1987). A view of the robotic grasping problem more related to biological research and artificial intelligence goals and techniques was developed in the early nineties (Venkataraman and Iberall 1990; Stansfield 1991; Bekey et al. 1993). Interdisciplinary research became more common, and works such as Mackenzie and Iberall (1994), with a real interdisciplinary stance, is still among the most cited in robot grasping. Nevertheless, engineering issues and approaches clearly keeps dominating the field (Shimoga 1996), and the currently most cited reviews on the subject, Bicchi (2000) and Okamura et al. (2000), are purely technical.

The survey of Bohg et al. (2014) reviews the most recents advances on grasp planning using data-driven methods. Grasping techniques are classified depending

E. Chinellato and A.P. del Pobil, *The Visual Neuroscience of Robotic Grasping*,
Cognitive Systems Monographs 28, DOI 10.1007/978-3-319-20303-4_3

on whether the target objects are previoulsy known, unknown or similar to a familiar category. Up to six different functional architectures are identified for organizing and structuring the necessary parts of a grasping system. For what concerns the planning of grasping actions through the use of visual information, a number of simplifications have commonly been used in robotics. The most common approach is to start from a model of the object to grasp, obtained or defined in an off-line stage, and to perform object recognition if necessary. A *grasp synthesis* process (see box below) can be thus performed on the model (Lopez-Damian et al. 2005), or pre-defined grasp programs can be accessed (Taylor and Kleeman 2004). Felip et al. (2013) propose a functional organization of grasping which relies on the concept of manipulation primitive. A primitive is an independent controller focused on a specific manipulation action. Primitives are coordinated through automatas which implement complete manipulation tasks. The same plan can be used on different hardware embodiments to execute the same task.

In the robotic literature (Bicchi 2000), a *grasp* is usually defined as the set of locations (points or regions) on an object surface where the effector—a simple gripper or an artificial hand—has to contact the object for grasping it. A grasp definition may include the configuration of the hand, which depends on its kinematics and *degrees of freedom* (DOF). *Grasp synthesis* is the problem of determining a proper set of contacts on the target object, and a corresponding suitable hand configuration. See Sahbani et al. (2012) for a review of grasp synthesis techniques. The inverse problem, *grasp analysis*, involves the study and evaluation of a given grasp.

The object model can be approximated with a set of shape primitives (Miller et al. 2003), or a 3D representation useful for grasp synthesis can be built performing a visual reconstruction of the object. Due to the implicit complexity of the task (Chaumette et al. 1996), visual reconstruction is often performed using range data (Ade et al. 1995; Rutishauser and Stricker 1995). As in these works, grasp synthesis often consists of searching for antipodal regions on the reconstructed object surfaces. Otherwise, the 3D model can be decomposed into basic structural components and grasps on object parts can be either generated (Goldfeder et al. 2007) or retrieved, according to predefined preshape primitives (Miller et al. 2003). Works that try to bias visual analysis with grasping aspects, without relying on models, have often dealt with simplified worlds, such as sets of planar objects (Stanley et al. 2000; Morales et al. 2006).

Only a few ambitious works have ever been developed for visual inspection of real 3D objects aimed at grasping actions. Taylor et al. (1994) deal with the problem of how a possible grip changes with the viewpoint, whereas Cipolla and Hollinghurst (1997) look for planar surfaces within a set of possible target objects and perform a grasping action with a parallel gripper. In Seitz (1999) a visual system is introduced that moves around a target object and is able to identify its major axis and extract a symbolic representation of the contour. On such representation, suitable grasping features, both parallel and curved, are searched for. The system of Saxena et al. (2008) learns to match visual object features to suitable grasping postures, showing a high adaptability and generalization capabilities in grasping with a parallel jaw-gripper. Preliminary results of the same method in which grasping experiments are executed with a three-finger Barrett Hand are also provided, and show that the technique is promising but need adaptation to the hand kinematics. Industrial applications of autonomous vision-based grasping, such as Sanz et al. (2005), are very rare. For the cases in which a set of candidate grasps are generated, quality measures focused on aspects related to the object and effector geometry and kinematics are often used in order to assess and select the most reliable and stable grasp configurations (Markenscoff et al. 1900; Ferrari and Canny 1992; Ponce and Faverjon 1995; Xiong et al. 1999; Borst et al. 2004; Chinellato et al. 2005).

Visual control of reaching and grasping movements and of manipulation actions is a very active area, often referred to as "visual servoing". The basic principle is that the movement follows a vision-based control law, with respect to either the Cartesian space or the image space (Hutchinson et al. 1996; Cervera et al. 2003; Recatalá et al. 2008a). These methods are not focused on the visual analysis of the object, but rather on the transport action and the fine adjustment of the movement. Such detailed visual tracking of the arm and hand movements is usually not performed by primates, unless in the case of sudden changes in the environment, or for very fine manipulation tasks. The most common solution in nature is the use of ballistic movements supported by extremely reliable proprioceptive and tactile information.

In fact, the use of visual information alone, without touch feedback, is as limiting in robotics as it is in primates. Nowadays, many works that deal not only with the synthesis but also with the execution of grasping actions, make use of force or tactile sensors on the fingers of the robot hand (Hollerbach 2000; Tegin and Wikander 2005). For instance, in Platt et al. (2002), force-based controllers are concurrently used to find an appropriate placement of the fingers on the object, and in Natale and Torres-Jara (2006) a robotic hand performs tactile exploration of the object, based on a set of exploratory primitives, in order to find a good grasp. More rarely, tactile information is used to recognize an object (Allen and Roberts 1989; Petriu et al. 2004) while active haptic exploration for detecting and modeling object features is practically a novel area in robotics (Petriu et al. 1992; Johnsson and Balkenius 2006). In any case, multimodal integration is a rapidly developing topic both at the practical and at the theoretical level (Pouget et al. 2002; Prats et al. 2013), and represents a fundamental issue for the future of grasping (Coelho et al. 2001; Lippiello et al. 2006; Grzyb et al. 2008; Prats et al. 2009) and robotic research in general (Barakova and Lourens 2005; Wermter et al. 2005).

3.2 Biological Inspiration for Robot Grasping and Manipulation

Many international research projects and meetings around the world are nowadays devoted to the interplay between robotics and life sciences (Dario et al. 2005). The use of robotic solutions for medical applications is the most developed and promising direction of such interdisciplinary approach, from aided microsurgery to prosthetics. Regarding upper limb mobility and grasping, brain guidance of artificial arms and hands has been performed with monkeys (Carmena et al. 2003), and the technology is being developed in order to achieve the same for humans (Andersen et al. 2005; Acharya et al. 2007). Pursuing this goal, very complex prosthetic hands are being constructed which join the dexterity of the most advanced robotic hands with new material technology for providing them with the best comfort and aspect (Buterfass et al. 2001; Huang et al. 2006; Cipriani et al. 2008). A review of the field, completed by a biomechanical comparison between the human hand and state of the art robotic manipulators provides a valuable framework for the study of the grasp both in nature and engineering (Leon et al. 2014).

For what concerns another aspect of the interplay between robotics and neuroscience—such as the use of insights regarding brain functions and natural solutions in general in order to implement better robotic systems—biologically inspired robotics is a rapidly developing field (Bar-Cohen and Breazeal 2003; Habib et al. 2007). Nevertheless, the limited literature about biological inspiration in visual-based robot grasping at the functional level suggests that the subject has still much to offer. In fact, for most biology-inspired grasping research available in the literature, the link with neuroscience is usually a general inspiration with limited impact on the final implementation (Kragic and Christensen 2003; Laschi et al. 2006). On the opposite end are works that appear as more biologically plausible, but which relation to life sciences is not clarified. For example, Kamon et al. (1998) propose a quite flexible and adaptable learning framework for grasping, but do not explicitly cite or mention any neuroscience research or biological inspiration. Systems that do implement neural mechanisms, such as Hoffmann et al. (2005), are rare, and work only with simplified environments and ad-hoc conditions.

In the development of systems which model or imitate natural skills, either cognitive or practical, two approaches are usually cited as contrary options: bottom-up and top-down. Modern robotics favors the former, in which complex behaviors emerge from simple, hierarchically organized ones (Brooks 1999). According to the paradigm of behavior-based robotics (Arkin 1998), complex tasks can be executed through the composition of simpler ones in a bottom-up direction, to the extent that some basic components are simple stimulus-effect reflexes of the agent with the environment (Braitenberg 1984).

Behavior-based grasping and manipulation robotic research is expanding (Zöllner et al. 2005; Recatalá et al. 2008b), but the paradigm is less widespread in grasping than it is in other areas of robotics. In fact, although some kinds of problems such as basic navigation of obstacle avoidance can be solved with simple and elegant behavior-based solutions, the high complexity of the vision, motor and especially the sensorimotor processing behind grasping actions are hard to be dealt with following such techniques.

While behavior-based robotics seems to follow the right direction toward mimicking biological systems, epigenetic robotics takes one step further in approaching natural intelligence, trying to evolve high-level abilities from basic ones as in an infant developing process (Berthouze and Goldfield 2008). This developmental process can be catalyzed by imitation or human teaching (Schaal 1999; Billard and Mataric 2000; Maistros and Hayes 2000; Ito et al. 2006). For what concerns manipulation, if not proper grasping, "baby" robots which learn simple manipulation skills from exploration and observation of their actions have been already built following this trend (Metta and Fitzpatrick 2003; Natale et al. 2005), and a possible future development is imitation between robots, which could accelerate the learning process without the need of humans to intervene. Some imitation processes used for robotic systems are also biologically plausible (Demiris 2002), as they are inspired by the mirror system of the primate brain, introduced in Sect. 2.3.3, and by the concept of *motor primitives* (see box).

Motor primitives are basic motor components used to generate more complex behaviors in all kinds of movements. The composition and functionality of the motor cortex described in Sect. 2.3.3 supports their use in primate visuomotor transformations. Motor primitives can be extracted from the thorough analysis of human motion (Drumwright et al. 2004), and used to form a motor vocabulary that can be employed to produce complex movements and action sequences (Nori and Frezza 2004).

Similarly to what happens in animals, the composition of simple behaviors as motor primitives can endow robotic systems with notable skills, dexterous manipulation being one of them (Mataric 2000). On this line, Kyota et al. (2005) use artificial neural networks to match simple voxel-based object representations to grasp configurations derived from basic human hand postures learnt with a data glove.

Even though the bottom-up approach has been usually considered more plausible from a physiological point of view, backprojections from associative to primary areas are widespread in the brain. Examples of bidirectional links, such as the premotor-parietal circuit, or the recurrent connections between visual areas, were given in the previous chapter. A mixed approach in which bottom-up mechanisms are triggered by top-down, cognitive style information, is therefore more faithful to the neurological reality. In artificial systems, whilst top-down solutions are implemented following knowledge engineering methodologies, bottom-up approaches are often coded with connectionist methods, more or less inspired by biological neural networks. Nowadays—and until artificial neural systems will resemble more closely the natural networks of neurons—more intelligent systems are probably better achieved with mixed approaches, in which pre-wired knowledge complements the connectionist mechanisms.

With the goal of linking bottom-up approaches with top-down cognitive representations, one of the most important issues is that of associating consistent symbolic meanings to aspects of the environment or of the agent-environment interaction. This issue is commonly known in cognitive science research as the *symbol grounding problem*. Though proposals have been put forth to solve the symbol grounding problem through robotic sensorimotor interactions, only little progress has been achieved with actual working systems. In the next section, a possible solution to the problem based on manipulation and grasping research is provided. Such approach can be useful in order to endow a robotic system with an implicit ability in merging practice with conceptual reasoning and thus "interpreting" its actions.

3.3 Symbol Grounding Through Robotic Manipulation

In this section, the problem of symbol grounding is addressed in the context of robotic manipulation. The goal is to obtain a more natural integration between the top-down and the bottom-up approaches by showing that there are symbols which do not refer simply to physical objects, but rather to the interactions between the robot and the objects in its environment. The description of two grasping and manipulation experiments performed at the Robotic Intelligence Lab of Universitat Jaume I, and summarized in this section, serves as a base on which to build a theory of symbolic representations for physical interactions. It will be shown that the symbols related to the interaction between agent and target object can be directly inferred from the execution of a planned action, and that neural networks can provide a suitable method for mapping complex perceptual signals to symbols. The proposal is sustained by important neuroscience studies, especially those related to the two streams of the visual cortex and the mirror system. Implementation details and further considerations can be found in Chinellato et al. (2007).

3.3.1 Symbol Grounding and Neuroscience

The symbol grounding problem is a classical challenge for cognitive science (Harnad 1990). In a traditional AI system, symbol interpretation is not intrinsic to the system, and the meaning of a symbol is always given by an external interpreter (e.g. the designer of the system). A really "intelligent" system should be able to assign symbolic meaning to an object or an action without being thought, as human beings can normally do. Harnad (1995) and del Pobil (1998) suggest that robotic sensorimotor interactions can represent a solution to this problem. In a cognitive robotic system, the symbols could be grounded in the system own capacity to interact physically with its environment.

So far, progress in symbol grounding by means of actual working robotic systems has been mostly related to visual recognition of individual objects. There is nevertheless an additional class of symbols, fundamental for cognitive robotics, which do not refer simply to physical objects but rather to the embodied physical interactions between the robot itself and the objects in the world. This kind of symbols would be more related to sensorimotor transformations, and they seem to have appeared in the evolutionary landscape long before vision as "sight" (Goodale and Westwood 2004).

The strategies employed by the brain when we interact with the world, and the more or less explicit symbolic meanings assigned to such interactions are an insightful source of inspiration for dealing with the symbol grounding problem in artificial agents. As explained above, motor primitives show different levels of complexity, and compose hierarchically to form a motor vocabulary. Complex movements and action sequences are composed in an almost linguistic way from this motor vocabulary. The question is whether motor and visuomotor primitives can be considered as symbols, extending the symbol grounding problem to a larger domain.

As explained in Chap. 2, visual processes related to specific actions in primates are different from visual processes not explicitly oriented to interaction of the subject with the environment. Area AIP codifies visual information in a grasp oriented way, associating visual features of a target object with a specific joint configuration suitable to guide the movements for grasping them, movements that are stored in the premotor cortex (Sakata et al. 2005). According to recent findings, in the inferior parietal lobule, and very likely even in AIP itself, actions are coded not only in a pragmatic way, but also in a semantic one (Gallese 2007). The activation of the dorsal stream areas of the mirror system, both parietal and premotor, when a subject is looking at an object with the purpose of interacting with it (e.g. reaching, hitting, pushing, grasping) seems to represent a "potential action". The emerging relation between sensory information and motor response may thus represent a symbolic correspondent of a grasping action (Hamilton and Grafton 2006). The dorsal stream involvement in action recognition suggests that actions are indeed associated with a symbolic meaning (Culham and Valyear 2006).

The lessons that robotics researchers can draw from these findings are: (1) the plausibility of a "dorsal style" visual elaboration that is exclusively dedicated to acting purposes; (2) the need of an emergent attribution of semantic meaning to synthesized grips. The next section introduces a robot system capable of object grasping in which visual analysis is dedicated to action, as in the primate dorsal stream, and in which a grasp codifies a relation between the hand and the object, as in humans.

3.3.2 An Emerging Categorization of Synthesized Robot Grips

Within the context of a robotic application for grasping and manipulation in a semi-structured environment, a framework was defined for characterizing candidate grips in a natural way, according to their properties in relation with the execution of a grasping action. Clustering of the candidate grip configurations occurs due to implicit

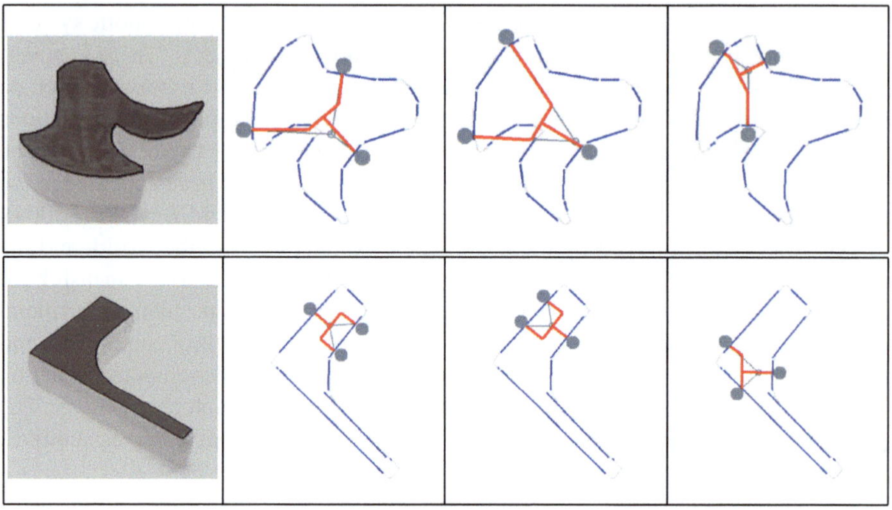

Fig. 3.1 Examples of three hand configurations found for two different objects

properties of grips, and an eventual symbolic meaning emerges through interaction of the agent with its world. The experimental setup used for the experiments described here is the one defined in Sect. 5.4.1.

The grasping action can be subdivided in four main modules or steps:

1. **Image processing:** The stereo vision system of the robot estimates the two-dimensional location of an unknown planar object placed on the table, which is the grasping target. A monocular image of the object is provided and analyzed to extract the object contour and identify on it regions suitable for finger contact.
2. **Grasp synthesis:** Several feasible candidate grasps (see some examples in Fig. 3.1) are generated, by selecting the target contact points for each triplet of grasp regions. Hand configurations for reaching the target points are computed taking into account the kinematic and geometric constraints of the Barrett Hand.
3. **Grasp evaluation:** The candidate grasps are characterized according to properties related to the geometry of the target object, the kinematics of the hand and their interaction, in order to perform a reasoned selection of the grasp to execute.
4. **Execution:** The hand is preshaped and positioned above the object, it moves down, closes the fingers so that the object is grasped, lifted and transported to a designated location. All this is performed with support of visual and tactile feedback.

Regarding grasp evaluation (step 3), a characterization scheme which includes nine high-level features was defined for providing an object-independent way to describe candidate grasps. In this way, each grasp is represented by nine measurements and thus by a point in a *9-dimensional* space. The features regard visual and motor aspects related to properties of the object contour, hand kinematics and nature of the contacts. The nine features derive from a set of quality criteria defined for assessing candidate grasp configurations (Chinellato et al. 2005).

A practical measurement of the reliability of a grasp was defined to characterize candidate configurations. After selecting a candidate grasp to execute, if the robot has been able to lift the object safely, a set of stability tests are applied in sequence. They consist of three consecutive shaking movements of the hand which are executed with an increasing acceleration. After each movement the tactile sensors are used to check whether the object has been dropped off. In this way the stability of the current grasp is measured.

An extensive experimental data gathering was realized following this protocol, and provided a qualitative measure of the success of many different grasps. A learning schema has been applied that makes use of the hyperspace described by the nine characterizing features, and tries to predict the reliability of a novel grip exploiting the information previously gathered on the outcome of already executed grasping actions. Different predicting methods have been applied and compared, with quite satisfying performances (Chinellato et al. 2003; Morales et al. 2004). These results showed that grasp reliability could be implicitly related to the characterizing features.

Summarizing, in this example each grip is represented as a point in a multidimensional space, and a procedure is introduced to predict a query point based on its similarity to previous grasping experiences. This is a case of *instance-based*, also known as *memory-based* learning (Aha 1997), which is a numeric version of the explicitly symbolic *case-based reasoning* (Waltz 1995). Although in this approach there is no explicit representation of the target function when training samples are provided, a natural characterization of the grips evolve from experience. The "character" of a new grip is recognized from the system as similar to some experience it already passed through, and the validity of such emerging associating ability is proved by the consequent predicting capacity of the system.

The system thus assigns each novel grip a symbolic value originating directly from its visuomotor character. Also, the visual analysis performed is only oriented toward action related features, as in the cortical dorsal stream, and there is no recognition of the target object. In this way, symbols are not related to the identity of objects, but rather to the different sensory experiences that objects can provide.

The next section proposes a procedure for extracting symbolic meaning from sensorimotor interactions through the use of connectionist methods.

3.3.3 Extracting Symbolic Meanings from Physical Interactions

In the previous section, a symbolic representation of sensorimotor experience was used to compare grips and predict the reliability of upcoming actions. This section describes an approach for extracting explicit symbolic meaning from sensor data

Fig. 3.2 System setup for the peg-in-hole task

through the use of artificial neural networks. In this case, the symbolic relation refers to the contact states between an already grasped object and the goal position.

The study is based on the classical two-dimensional peg-in-hole insertion task (Cervera and del Pobil 2000), performed using the setup depicted in Fig. 3.2. The task is to insert a peg hold with the gripper in a chamferless hole. Wrist-mounted force sensors are used to measure the external forces and torques applied to the peg as the manipulator interacts with the environment.

Symbol grounding in this experiment is required to assign symbolic values to the physical interactions between the robot—the peg can be considered as a prolongation of the robot gripper—and the environment. The correct identification of such interactions is of fundamental importance for the adequate execution of the task: in this particular case the insertion of the peg into the hole. Physical interactions are modeled as contact states between peg and hole, that have to be identified with the help of the force sensors. In Fig. 3.3, all possible contact states and the corresponding forces are shown. The no-contact state, **F0**, shows the weight force, while all the others only show the reaction forces, but weight is considered too. In order to identify the contact states, only the direction of forces is relevant, not their magnitude. The sensors provide only three raw signals of force and torque, and the problem is to

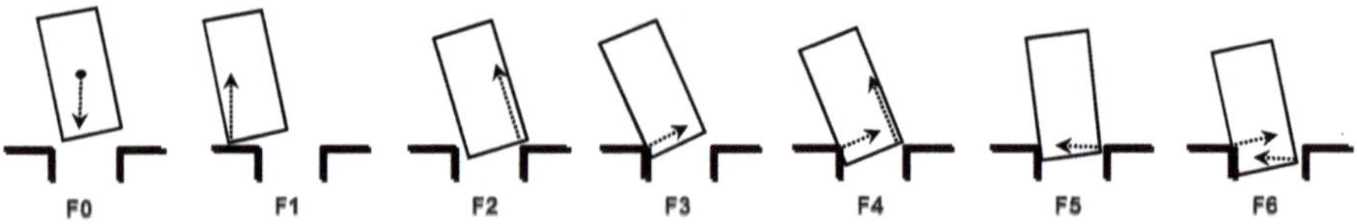

Fig. 3.3 Set of contact states for the peg-in-hole task, showing the contact forces between peg and hole. State **F0** shows only the weight force which is omitted in all other graphs

Table 3.1 Contact states classification percentages

State	Right	Wrong	Unknown
F1	94	5	1
F2	100	–	–
F3	59	34	7
F4	97	–	3
F5	100	–	–
F6	78	21	1
Average	88	10	2

appropriately map these signals to one symbolic contact state. This is not a trivial problem due to the variability of the forces and the superposition of the peg weight and one or several reaction forces.

In order to perform the matching between sensory data and symbolic states, Self-Organizing Maps were used (Kohonen 1990). The experiments consist of three phases:

1. **Training.** The neural network is trained with a set of 1200 random input samples equally distributed across contact states, according to the procedure described in Kohonen (1990). In this way the net learns in an unsupervised way the natural structure of the data.
2. **Calibration.** A set of 600 labeled samples is used for calibrating the net to the contact states. The network response is analyzed for all inputs, and each network unit is labeled with the contact state for which it has been more frequently selected as the closest unit.
3. **Testing.** The performance of the network is tested with an independent set of 600 samples. For each sample, the most responsive unit is selected, and the output contact state is given by that unit's label. An uncertain response occurs if the unit is unlabeled.

Results from a typical experiment are shown in Table 3.1. Two states, **F2** and **F5**, are perfectly identified. State **F4** is correctly identified in the 97 % of cases and **F1** is properly classified in the 94 % of the cases. States **F3** and **F6** seem to be more difficult to identify, and the correct classification percentages are smaller. The average network performance, 88 % success, is anyway very good, and it has to be taken into account that only force information has been used.

After establishing the correspondence between force sensor data and symbols denoting contacts, the peg-in-hole insertion problem can be solved in a number of ways. These are depicted in Fig. 3.4, that shows a perception-based diagram in which the increments inside the nodes denote the actions to be performed for reaching the

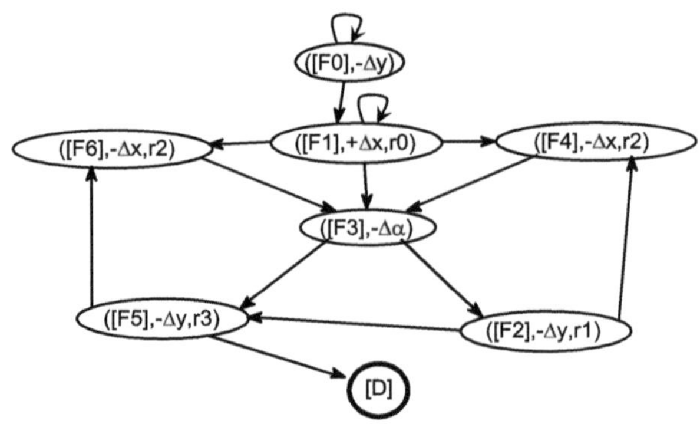

Fig. 3.4 Perception-based motion plan for solving the peg-in-hole task. The increments within the nodes symbolize the possible actions and the arrows the consequent state transitions

neighbor nodes. The transitions are defined so that the goal position **D** can be reached from any state. Several tests were performed with a real robot and sensors using the transition rules of Fig. 3.4, and following different sequences of contact states. The good experimental results demonstrated that the task could actually be performed by using direct sensory data only, without the need for information about the position and orientation of the peg.

This example shows how self-organizing neural networks were able to automatically extract a number of symbolic states that represent relevant real-world situations useful in a transition graph for the solution of a relatively complex problem. Most importantly, such symbolic knowledge derived directly from sensory information regarding ongoing actions.

3.3.4 Symbolic Value of Hand-Object Interactions

The two case studies exposed above show how symbolic meanings can naturally arise from the physical interaction between an agent and objects in its environment, suggesting that motor primitives constitute a consistent source of symbolic knowledge.

Looking once more at cut-edge research in neurophysiology, the existence of a parieto-premotor mirror system (Sect. 2.5) supports the idea of extending the symbol concept to motor behaviors and sensorimotor interactions. In fact, the symbols managed by AIP do not codify objects, but action-oriented visual representations, and association patterns between objects and distal subject effectors (Tunik et al. 2007). Taking a step further in the analysis of the mirror mechanisms, it has been proposed that symbolic communication and language evolved from a neural motor system involved in action recognition (Keysers et al. 2003; Hamilton and Grafton 2006). Moreover, the mirror system seems to play a critical role in learning by imitation, a skill that we are only beginning to develop in robots. Broca's area in humans, traditionally related to language production, is the most likely correspondent of F5, where mirror neurons were first discovered (Binkofski and Buccino 2004; Grèzes et al. 2003). This would confirm that action understanding, recognition and mental imagery of actions do not differ conceptually from object recognition or imaging, and that complex cognitive processes emerge from simple behaviors which firstly evolved in order to endow the organism with skills for better interacting in its environment.

3.4 Toward Intelligent Robotic Grasping

In the previous section, the existence of symbols that are grounded in the physical sensorimotor interactions between the robot and the objects in the world has been discussed. The two presented case studies show how symbolic meaning can describe, and be extracted from, visuomotor experiences regarding manipulation tasks, allowing the robot to build a representation of the possible interactions with its surrounding environment. The merging of different sensory modalities, such as vision and touch, in the exploration of the environment is probably the most necessary further development of the proposed approach. A fundamental conclusion, consistent with findings about the action-oriented human dorsal visuomotor stream, is that in a robotic system there is no need to model, recognize or classify an object in order to physically interact with it, since the symbols derived from sensorimotor experience identify particular physical interactions between the robot hand and the target object.

As previously explained, classical robotic approaches do not follow this path. Instead, they consider either object recognition or full visual reconstruction previous to grasp analysis, which is normally based on an object model (Bicchi 2000). This pattern corresponds to the scheme of Fig. 3.5a, in which perception-synthesis-action

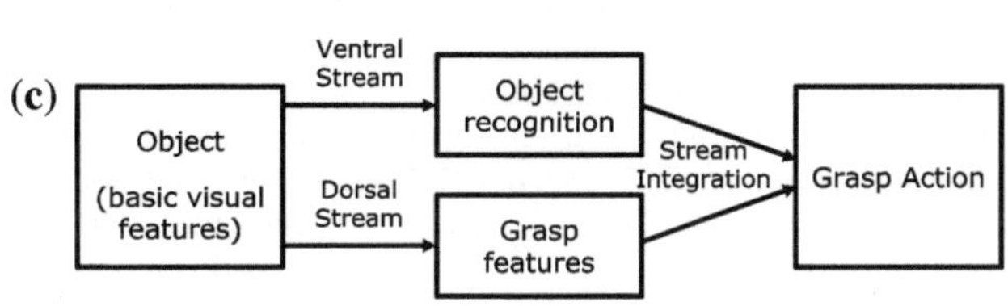

Fig. 3.5 Different approaches in vision-based grasping: **a** is the traditional perception-reason-act paradigm; **b** represents an action-based vision approach; **c** is the integrated methodology proposed in this work

are sequentially executed, and vision is general purpose rather than goal-oriented. There are works in the literature which exploit visual features in a manner focused on grasp purposes, but without the aid of "cognitive" information about objects (see e.g. Morales et al. 2006; Saxena et al. 2008 and Sect. 3.3.2). This approach is symbolized in Fig. 3.5b. The first method lacks of flexibility, as it builds on a nearly general-purpose visual processing stage. The second method partially resembles the job of the dorsal stream, and is thus more biologically plausible and often very efficient. Nevertheless, it does not exploit the kind of cognitive information that makes human grasping skills largely transcend a geometric or kinematic analysis.

The review presented in this chapter shows that the grasp problem in robotics is still largely unsolved. The proposal put forth in this book for obtaining a more reliable and meaningful interaction with real world objects is to integrate in a robotic grasping system instantaneous visual information gathered in an action oriented-manner (dorsal pathway) with experience-mediated information (ventral pathway).

In Fig. 3.5c a representation of this solution, which somehow encompasses the two previous ones, is depicted. This integration process is what human beings do all the time: we join our knowledge about the objects we are going to grasp, and the experience of previous actions, with the analysis of the actual, concrete situation we are facing. The model introduced in the next chapter goes toward this direction, and analyzes the indications of neuroscience research from a practical point of view, in order to devise a set of mechanisms able to endow a robot grasping system with increased capabilities in interacting with its nearby environment.

References

Acharya S, Aggarwal V, Tenore F, Shin HC, Etienne-Cummings R, Schieber M, Thakor N (2007) Towards a brain-computer interface for dexterous control of a multi-fingered prosthetic hand. In: International IEEE/EMBS conference on neural engineering, pp 200–203. doi:10.1109/CNE.2007.369646

Ade F, Rutishauser M, Trobina M (1995) Grasping unknown objects. In: Bunke H, (ed) Dagstuhl proceedings seminar. World Scientific, pp 445–459

Aha D (1997) Lazy learning. Artif Intell Rev 11:7–10

Allen P, Roberts K (1989) Haptic object recognition using a multi-fingered dextrous hand. In: IEEE international conference on robotics and automation, pp 342–347. doi:10.1109/ROBOT.1989.100011

Andersen R, Andersen R, Musallam S, Burdick J, Cham J (2005) Cognitive based neural prosthetics. In: IEEE international conference on robotics and automation, pp 1908–1913

Arkin R (1998) Behavior-based robotics. MIT Press, Cambridge

Barakova EI, Lourens T (2005) Event based self-supervised temporal integration for multimodal sensor data. J Integr Neurosci 4(2):265–282. doi:10.1142/S021963520500077X

Bar-Cohen Y, Breazeal C (2003) Biologically inspired intelligent robots. SPIE Press, San Diego

Bekey G, Liu H, Tomovic R, Karplus W (1993) Knowledge-based control of grasping in robot hands using heuristics from human motor skills. IEEE Trans Robot Autom 9(6):709–722

Berthouze L, Goldfield EC (2008) Assembly, tuning, and transfer of action systems in infants and robots. Infant Child Dev 17(1):25–42. doi:10.1002/icd.542

Bicchi A (2000) Hand for dexterous manipulation and robust grasping: a difficult road towards simplicity. IEEE Trans Robot Autom 16(6):652–662

Billard A, Mataric M (2000) Learning motor skills by imitation: a biologically inspired robotic model. In: International conference on cognitive and neural systems, Boston

Binkofski F, Buccino G (2004) Motor functions of the Broca's region. Brain Lang 89(2):362–369. doi:10.1016/S0093-934X(03)00358-4

Bohg J, Morales A, Asfour T, Kragic D (2014) Data-driven grasps synthesis—a survey. IEEE Trans Robot 30(2):289–309

Borst C, Fischer M, Hirzinger G (2004) Grasp planning: how to choose a suitable task space. In: IEEE international conference on robotics and automation, New Orleans, USA, pp 319–325

Braitenberg V (1984) Vehicles: experiments in synthetic psychology. MIT Press, Cambridge

Brooks R (1999) Cambrian intelligence: the early history of the new AI. MIT Press, Cambridge

Buterfass J, Grebenstain M, HLiu, Hirzinger G (2001) DLR-Hand II: next generation of a dextrous robot hand. In: IEEE international conference on robotics and automation, Seoul, Korea

Carmena JM, Lebedev MA, Crist RE, O'Doherty JE, Santucci DM, Dimitrov DF, Patil PG, Henriquez CS, Nicolelis MAL (2003) Learning to control a brain-machine interface for reaching and grasping by primates. PLoS Biol 1(2):e42. doi:10.1371/journal.pbio.0000042

Cervera E, Pobil APD, Berry F, Martinet P (2003) Improving image-based visual servoing with three-dimensional features. Int J Rob Res 22(10–11):821–840

Cervera E, del Pobil A (2000) A qualitative-connectionist approach to robotic spatial planning. Spat Cognit Comput 1(2):51–76

Chaumette F, Boukir S, Bouthemy P, Juvin D (1996) Structure from controlled motion. IEEE Trans Pattern Anal Mach Intell 18(5):492–504

Chinellato E, Morales A, Fisher RB, del Pobil AP (2005) Visual quality measures for characterizing planar robot grasps. IEEE Trans Syst Man Cybern Part C 35(1):30–41

Chinellato E, Morales A, Cervera E, del Pobil AP (2007) Symbol grounding through robotic manipulation in cognitive systems. Robot Auton Syst 55(12):851–859

Chinellato E, Morales A, Sanz PJ, del Pobil AP (2003) Validation of features for characterizing robot grasps. In: Mira J, Alvarez J (eds) Artificial neural nets problem solving methods, LNCS 2687. Springer, pp 193–200

Cipolla R, Hollinghurst N (1997) Visually guided grasping in unstructured environments. Robot Auton Syst 19:337–346

Cipriani C, Zaccone F, Micera S, Carrozza MM (2008) On the shared control of an EMG-controlled prosthetic hand: analysis of user-prosthesis interaction. IEEE Trans Robot Autom 24(1):170–184. doi:10.1109/TRO.2007.910708

Coelho J Jr, Pieter J, Grupen R (2001) Developing haptic and visual perceptual categories for reaching and grasping with a humanoid robot. Robot Auton Syst 37:195–218

Culham JC, Valyear KF (2006) Human parietal cortex in action. Curr Opin Neurobiol 16(2):205–212. doi:10.1016/j.conb.2006.03.005

Cutkosky M (1985) Robotic grasping and fine manipulation. Kluwer Academic Press, Dordrecht

Dario P, Carrozza M, Guglielmelli E, Laschi C, Menciassi A, Micera S, Vecchi F (2005) Robotics as a future and emerging technology: biomimetics, cybernetics, and neuro-robotics in European projects. IEEE Robot Autom Mag 12(2):29–45. doi:10.1109/MRA.2005.1458320

del Pobil AP (1998) The grand challenge is called: robotic intelligence. In: Tasks and methods in applied artificial intelligence, LNCS1416. Springer, pp 15–24

Demiris Y (2002) Mirror neurons, imitation, and the learning of movement sequences. In: International conference on neural information processing, Singapore, pp 111–115

Drumwright E, Jenkins O, Mataric M (2004) Exemplar-based primitives for humanoid movement classification and control. In: IEEE international conference on robotics and automation, New Orleans, USA

Felip J, Laaksonen J, Morales A, Kyrki V (2013) Manipulation primitives: a paradigm for abstraction and execution of grasping and manipulation tasks. Robot Auton Syst 61(3):283–296

Ferrari C, Canny J (1992) Planning optimal grasps. In: IEEE international conference on robotics and automation, Nice, France, pp 2290–2295

Gallese V (2007) The "conscious" dorsal stream: embodied simulation and its role in space and action conscious awareness. Psyche 13(1):

Goldfeder C, Allen P, Pelossof R, Lackner C (2007) Grasp planning via decomposition trees. In: IEEE international conference on robotics and automation, Rome, Italy

Goodale MA, Westwood DA (2004) An evolving view of duplex vision: separate but interacting cortical pathways for perception and action. Curr Opin Neurobiol 14(2):203–211. doi:10.1016/j.conb.2004.03.002

Grèzes J, Armony JL, Rowe J, Passingham RE (2003) Activations related to "mirror" and "canonical" neurones in the human brain: an fMRI study. Neuroimage 18(4):928–937

Grzyb BJ, Chinellato E, Morales A, del Pobil AP (2008) Robust grasping of 3D objects with stereo vision and tactile feedback. In: IEEE international conference on intelligent robots and systems

Habib MK, Watanabe K, Izumi K (2007) Biomimetics robots from bio-inspiration to implementation. In: Annual conference of the ieee industrial electronics society, pp 143–148. doi:10.1109/IECON.2007.4460382

Hamilton AF, Grafton ST (2006) Goal representation in human anterior intraparietal sulcus. J Neurosci 26(4):1133–1137. doi:10.1523/JNEUROSCI.4551-05.2006

Harnad S (1995) Grounding symbolic capacity in robotic capacity. In: Brooks R (ed) Steels. The artificial life route to artificial intelligence, building situated embodied agents, Lawrence Erlbaum, pp 277–286

Harnad S (1990) The symbol grounding problem. Phys D 42:335–346. doi:10.1016/0167-2789(90)90087-6

Hoffmann H, Schenck W, Möller R (2005) Learning visuomotor transformations for gaze-control and grasping. Biol Cybern 93(2):119–130. doi:10.1007/s00422-005-0575-x

Hollerbach JM (2000) Some current issues in haptics reseach. In: IEEE international conference on robotics and automation, San Francisco, USA, pp 757–762

Huang H, Li J, Liu Y, Hou L, Cai H, Liu H (2006) The mechanical design and experiments of HIT/DLR prosthetic hand. In: IEEE international conference on robotics and biomimetics, pp 896–901. doi:10.1109/ROBIO.2006.340339

Hutchinson S, Hager G, Corke P (1996) A tutorial on visual servo control. IEEE Trans Robot Autom 12(5):651–670

Iberall T (1987) The nature of human prehension: three dextrous hands in one. In: IEEE international conference on robotics and automation, pp 396–401

Ito M, Noda K, Hoshino Y, Tani J (2006) Dynamic and interactive generation of object handling behaviors by a small humanoid robot using a dynamic neural network model. Neural Netw 19(3):323–337. doi:10.1016/j.neunet.2006.02.007

Johnsson M, Balkenius C (2006) Experiments with artificial haptic perception in a robotic hand. J Intell Fuzzy Syst 17(4):377–385

Kamon I, Kamon I, Flash T, Edelman S (1998) Learning visually guided grasping: a test case in sensorimotor learning. IEEE Trans Syst Man Cybern Part A 28(3):266–276. doi:10.1109/3468.668958

Keysers C, Kohler E, Umiltà MA, Nanetti L, Fogassi L, Gallese V (2003) Audiovisual mirror neurons and action recognition. Exp Brain Res 153(4):628–636. doi:10.1007/s00221-003-1603-5

Kohonen T (1990) The self-organizing map. In: Proceedings of IEEE, vol 78, no 9, pp 1464–1480

Kragic D, Christensen HI (2003) Biologically motivated visual servoing and grasping for real world tasks. In: IEEE international conference on intelligent robots and systems, Las Vegas, USA

Kyota F, Watabe T, Saito S, Nakajima M (2005) Detection and evaluation of grasping positions for autonomous agents. In: International Workshop on Language Understanding and Agents for Real World Interaction, pp 453–460

Laschi C, Asuni G, Teti G, Carrozza M, Dario P, Guglielmelli E, Johansson R (2006) A bio-inspired neural sensory-motor coordination scheme for robot reaching and preshaping. In: IEEE international conference on biomedical robotics and biomechatronics, pp 531–536

Leon B, Morales A, Sancho-Bru J (2014) From robot to human grasping simulation, vol 19. Springer, New York

Lippiello V, Siciliano B, Villani L (2006) Robot interaction control using force and vision. In: IEEE international conference on intelligent robots and systems, pp 1470–1475. doi:10.1109/IROS.2006.281974

Lopez-Damian E, Sidobore D, Alami R (2005) Grasp planning for non-convex objects. In: International symposium on robotics, Tokyo, Japan

Mackenzie C, Iberall T (1994) The grasping hand. North Holland, New York

Maistros G, Hayes G (2000) An imitation mechanism inspired from neurophysiology. In: International workshop on current computational architectures integrating neural networks and neuroscience, Durham, UK

Markenscoff X, Li L, Papadimitriou C (1990) The geometry of grasping. Int J Robot Res 9(1):61–74

Mason M, Salisbury J Jr (1985) Robot hands and the mechanics of manipulation. MIT Press, Cambridge

Mataric MJ (2000) Getting humanoids to move and imitate. IEEE Intell Syst 15(4):18–24

Metta G, Fitzpatrick P (2003) Early integration of vision and manipulation. Adapt Behav 11(2):109–128

Miller AT, Knopp S, Christensen HI, Allen PK (2003) Automatic grasp planning using shape primitives. In: IEEE international conference on robotics and automation, Taipei, Taiwan

Morales A, Chinellato E, Fagg AH, del Pobil AP (2004) Using experience for assessing grasp reliability. Int J Humanoid Rob 1(4):671–691

Morales A, Sanz PJ, del Pobil AP, Fagg AH (2006) Vision-based three-finger grasp synthesis constrained by hand geometry. Robot Auton Syst 54(6):496–512

Napier J (1983) Hands. Princeton University Press, Princeton

Natale L, Metta G, Sandini G (2005) A developmental approach to grasping. In: Developmental robotics AAAI spring symposium

Natale L, Torres-Jara E (2006) A sensitive approach to grasping. In: International conference on epigenetic robotics, Paris, France

Nguyen VD (1988) Constructing force-closure grasps. Int J Robot Res 7(3):3–16

Nori F, Frezza R (2004) Biologically inspired control of a kinematic chain using the superposition of motion primitives. In: IEEE conference on decision and control, vol 1, pp 1075–1080

Okamura A, Smaby N, Cutkosky M (2000) An overview of dexterous manipulation. In: IEEE international conference on robotics and automation, San Francisco, CA, USA, pp 255–260

Petriu EM, McMath WS, Yeung SSK, Trif N (1992) Active tactile perception of object surface geometric profiles. IEEE Trans Instrum Measur 41(1):87–92

Petriu EM, Yeung SKS, Das SR, Cretu AM, Spoelder HJW (2004) Robotic tactile recognition of pseudorandom encoded objects. IEEE Trans Instrum Meas 53(5):1425–1432. doi:10.1109/TIM.2004.834100

Platt R, Fagg A, Grupen R (2002) Nullspace composition of control laws for grasping. In: IEEE international conference on intelligent robots and systems, Lausanne, Switzerland, pp 1717–1723

Ponce J, Faverjon B (1995) On computing three-finger force-closure grasps of polygonal objects. IEEE Trans Robot Autom 11(6):868–881

Pouget A, Deneve S, Duhamel JR (2002) A computational perspective on the neural basis of multisensory spatial representations. Nat Rev Neurosci 3(9):741–747. doi:10.1038/nrn914

Prats M, del Pobil AP, Sanz PJ (2013) Robot physical Interaction through the combination of Vision, Tactile and Force Feedback, vol 84. Springer, New York

Prats M, Sanz PJ, del Pobil A (2009) Vision-tactile-force integration and robot physical interaction. In: IEEE International Conference on Robotics and Automation, 2009, ICRA '09, pp 3975–3980. doi:10.1109/ROBOT.2009.5152515

Recatalá G, Carloni R, Melchiorri C, Sanz P, Cervera E, del Pobil A (2008a) Vision-based grasp tracking for planar objects. IEEE Trans Syst Man Cybern Part C Appl Rev 38(6):844–849

Recatalá G, Chinellato E, del Pobil AP, Mezouar Y, Martinet P (2008b) Biologically-inspired 3D grasp synthesis based on visual exploration. Auton Robots 25(1):59–70. doi:10.1007/s10514-008-9086-7

Rutishauser M, Stricker M (1995) Searching for grasping opportunities on unmodeled 3D objects. In: British machine vision conference, vol 1, pp 277–286

Sahbani A, El-Khoury S, Bidaud P (2012) An overview of 3d object grasp synthesis algorithms. Robot Auton Syst 60(3):326–336. doi:10.1016/j.robot.2011.07.016

Sakata H, Tsutsui KI, Taira M (2005) Toward an understanding of the neural processing for 3D shape perception. Neuropsychologia 43(2):151–161. doi:10.1016/j.neuropsychologia.2004.11.003

Sanz PJ, Requena A, Iñesta JM, Del Pobil AP (2005) Grasping the not-so-obvious: vision-based object handling for industrial applications. IEEE Robot Autom Mag 12(3):44–52. doi:10.1109/MRA.2005.1511868

Saxena A, Driemeyer J, Ng AY (2008) Robotic grasping of novel objects using vision. Int J Robot Res 27(2):157–173. doi:10.1177/0278364907087172

Schaal S (1999) Is imitation learning the route to humanoid robots? Trends Cogn Sci 3(6):233–242

Seitz M (1999) Towards autonomous robotic servicing: using an integrated arm-eye system for manipulating unknown objects. Robot Auton Syst 26(1):23–42

Shimoga K (1996) Robot grasp synthesis algorithms: a survey. Int J Robot Res 15(3):230–266

Stanley K, Wu J, Gruver W (2000) Implementation of vision-based planar grasp planning. IEEE Trans Syst Man Cybern 30(4):517–524

Stansfield S (1991) Robotic grasping of unknown objects: a knowledge-based approach. Int J Robot Res 10(4):314–326

Taylor M, Blake A, Cox A (1994) Visually guided grasping in 3D. In: IEEE international conference on robotics and automation, San Diego, California, pp 761–766

Taylor G, Kleeman L (2004) Integration of robust visual perception and control for a domestic humanoid robot. In: IEEE International conference on intelligent robots and systems, pp 1010–1015. doi:10.1109/IROS.2004.1389485

Tegin J, Wikander J (2005) Tactile sensing in intelligent robotic manipulation—a review. Ind Robo Int J 32:64–70. doi:10.1108/01439910510573318

Tunik E, Rice NJ, Hamilton A, Grafton ST (2007) Beyond grasping: representation of action in human anterior intraparietal sulcus. Neuroimage 36(Suppl 2):T77–T86. doi:10.1016/j.neuroimage.2007.03.026

Venkataraman S, Iberall T (eds) (1990) Dextrous robot hands. Springer, New York

Waltz DL (1995) Memory-based reasoning. In: Arbib MA (ed) The handbook of brain theory and neural networks, MIT Press, pp 661–662

Wermter S, Weber C, Elshaw M (2005) Associative neural models for biomimetic multi-modal learning in a mirror neuron-based robot. In: Cangelosi A, Bugmann G, Borisyuk R (eds) Modeling language cognition and action. World Scientific, pp 31–46

Xiong CH, Li YF, Ding H (1999) On the dynamic stability of grasping. Int J Robot Res 18(9):951–958

Zöllner R, Pardowitz M, Knoop S, Dillmann R (2005) Towards cognitive robots: building hierarchical task representations of manipulators from human demostration. In: IEEE international conference on robotics and automation, Barcelona, Spain, pp 1547–1552

Chapter 4
Vision-Based Grasping, Where Robotics Meets Neuroscience

In the previous chapters, the current knowledge on the neuroscience of vision-based grasping in humans and other primates was analyzed, with the goal of establishing the computational bases for a robotic system able to achieve advanced grasping skills in the real world. The way chosen to achieve such goal is through the integration of on-line, action-oriented visual information (dorsal pathway) with knowledge about the target object and memories of previous grasping experiences and object characteristics (ventral pathway).

Previous models of vision-based grasping have built so far mainly, when not exclusively, on monkey data. Recent neuropsychological and neuroimaging research has shed a new light on how visuomotor coordination is organized and performed in the human brain. Thanks to such research, a model of vision-based grasping which integrates knowledge coming from single-cell monkey studies with human data can be developed. The basic framework of the proposed model is outlined in this chapter. Final goal of the proposal is to mimic, in a robotic setup, the coordination between sensory, associative and motor cortex of the human brain in vision-based grasping actions.

In the following section, previous related models are reviewed. Then, basic issues regarding robotic and human grasping, their differences and similarities, and the way to link them through computational modeling are discussed. Next, the full outline of the model is presented.

4.1 Previous Models and Related Approaches

Computational modeling of visual mechanisms in mammals is a wide and developed area (Rolls and Deco 2002), that dates back to the sixties (Hubel and Wiesel 1962). Strongly biologically inspired models of visual areas have been implemented, focusing mainly on the primary visual cortex (see e.g. Sabatini et al. 2001; McLaughlin et al. 2003; Lourens and Barakova 2007), or on increasingly invariant object representations for recognition as performed along the ventral stream (Ullman 1996; O'Reilly

E. Chinellato and A.P. del Pobil, *The Visual Neuroscience of Robotic Grasping*,
Cognitive Systems Monographs 28, DOI 10.1007/978-3-319-20303-4_4

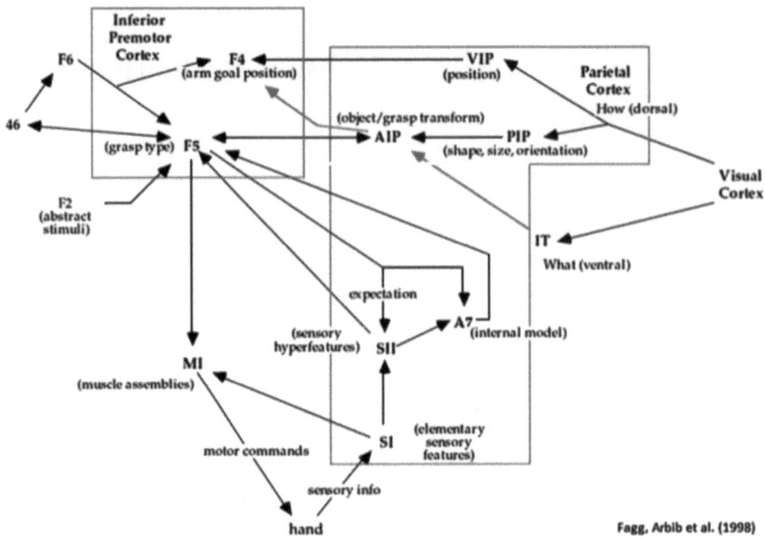

Fig. 4.1 Model framework of Fagg and Arbib (1998)

and Munakata 2000; Riesenhuber and Poggio 2000; Cadieu et al. 2007). Bio-inspired systems for object recognition (Khosla et al. 2014) and biological motion detection (Yousefi and Loo 2014a, b) are also available Complex agent-object interactions—such as in grasping actions—are not a usual target of computational models. The integration between the contributions of the two visual pathways is a subject almost unexplored (Rolls and Deco 2002; Rolls and Webb 2014), and only a few studies explicitly deal with dorsal stream processing.

The most complete attempt to computationally describe the sensorimotor mechanisms of visual-based grasping in primates is the FARS (Fagg-Arbib-Rizzolatti-Sakata) model (Fagg and Arbib 1998), whose framework is depicted in Fig. 4.1. The FARS model focuses especially on the interaction between AIP and F5, and is oriented to the action execution part of the process. The model is implemented using biologically inspired neural networks, and includes a large number of different brain areas (mainly inspired on monkey physiology), even though only areas AIP, F5 and the primary motor cortex F1 are modeled in detail. For AIP, the distinction between visual, visuomotor and motor neurons is taken into account, but the most recent classification of object and non-object type neurons (Sect. 2.3.2.1) is not.

The main achievement of the FARS model is the reproduction, at a qualitative level, of the variety of cell behaviors observed in F5 and AIP. Coordination in time between the two areas is well implemented: AIP assumes the role of action planner and working memory, while F5 composes action execution at a high level, the motor commands being elaborated by F1. In the FARS model, a task is given also to the inferior temporal cortex IT, the area of a monkey's brain in charge of object recognition: IT takes into account the object identity, and recall a related affordance

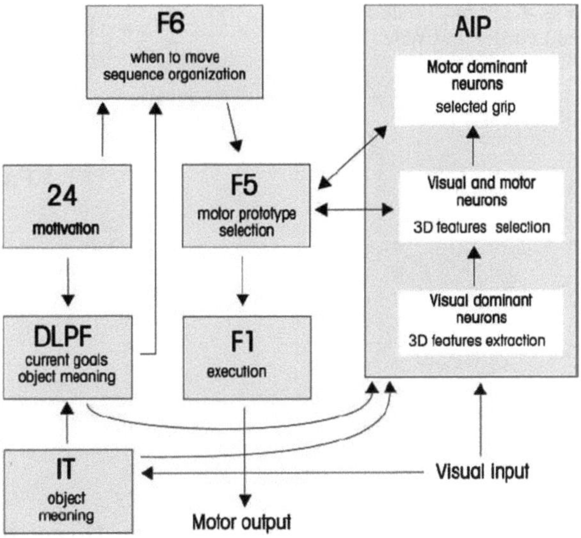

Fig. 4.2 Model framework of Rizzolatti and Luppino (2001)

if the association is available. This represents a plausible, although simple, modeling of the two streams interaction. Previous robotic research, developed as a collaboration between the Robotic Intelligence Lab of Universitat Jaume I and Andrew Fagg (at that time at the Laboratory for Perceptual Robotics of University of Massachusetts) was loosely inspired by the FARS model, but computational mechanisms were not thoroughly considered (Chinellato et al. 2003a; Morales et al. 2004, 2006). An interesting extension of the FARS model includes a mirror neuron module (Oztop and Arbib 2002), suggesting a role in the development of language for the two streams (Arbib 2010).

Rizzolatti and Luppino (2001) suggested that the FARS model should be modified, as the hypothesis that action selection is obtained in a loop F5-AIP, using information coming from the ventral stream and the prefrontal cortex, seems to be not coherent with more recent findings. According to the authors (see Fig. 4.2), AIP is the site of action selection, as it receives direct input from the ventral stream and the prefrontal cortex, whilst F5 does not. AIP would send the coordinates of only the selected affordance to F5, and the action would remain potential until a release signal is received.

Fukuda et al. (2000) propose a model for grasp synthesis from visual information based on observation of real human movements. The authors use a neural network to match sampled images to grasp configurations, concluding that it is possible to generate the proper hand preshaping relying only on information extracted from visual data. The different role of the two visual streams in the process is not taken into account. This work will be more throughly discussed in Sect. 6.2. Fukui et al. (2006) experiment and model on the control of preshaping, and get to similar conclusions. Although the work does not explicit deal with grasp planning, it is argued

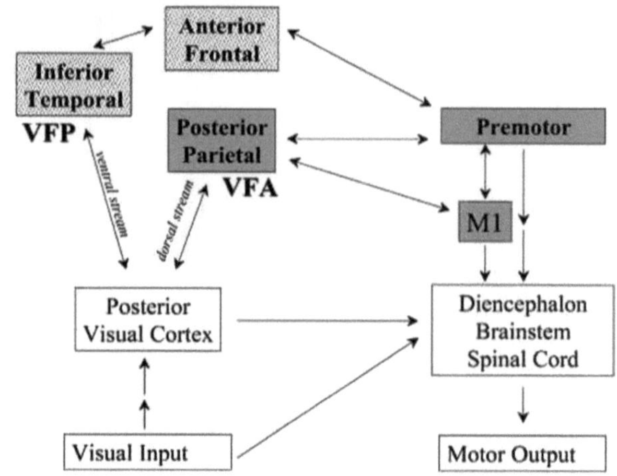

Fig. 4.3 Model framework of Lebedev and Wise (2002)

that preshaping and finger closing are related to purely visuomotor mechanisms, and driven by a control of predictive nature. Tactile input enters the process only when the object is touched, even though there is probably a prediction of the expected tactile feedback.

The model of Lebedev and Wise (2002) distinguishes between a vision for action (VFA) and a vision for perception (VFP) systems, that can be approximately identified with the dorsal and ventral streams respectively. According to the authors, the interaction between the systems takes place as a bidirectional process. The VFA system tries to develop the target action so as to better fulfill the contingent sensorial situation, and gets help from the VFP system which in turn biases the selection towards stored, recognizable patterns (Fig. 4.3). Dorsal stream areas also seem to participate in more perceptual processes. The model proposed in this book extends these assumptions, postulating a bidirectional connection between ventral and dorsal areas, as indicated by the most recent research on the two streams (Sect. 2.4.3). The basic idea is that not only the LOC helps AIP in the action selection process, but AIP and/or CIP support object recognition, providing the ventral stream with data on possible object affordances, and facilitating in this way the identification process. Interesting in Lebedev and Wise (2002) is also the role assigned to basal ganglia and prefrontal cortex, areas that affect the working of both streams. A drawback of the model is that no details for computational implementation of the proposed concepts are given.

The model of Zhonget al. (2014) is based on a Recurrent Neural Network with Parametric Bias units (RNNPB) and can run either in Learning, Recognition and Prediction mode (Fig. 4.4). It is exemplified through a robot passively observing an object to learn its features and movements. The information in the dorsal and ventral streams is separately encoded in two network streams, and the predictions of both streams are brought together via the horizontal product while the PB units act as a conceptualization of both streams. Even though no details on brain areas

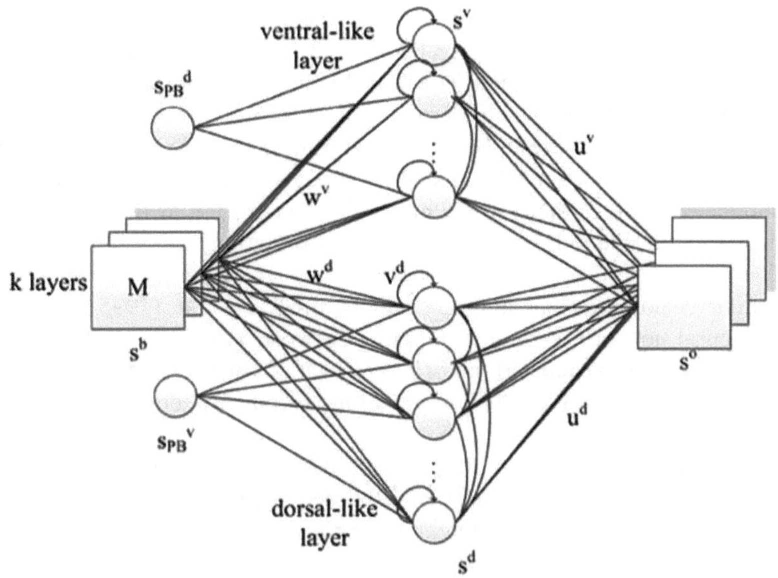

Fig. 4.4 Model framework of Zhonget al. (2014)

are considered, the model helps in shedding light on the feasibility of linking the conceptualization of the two streams and the predictive function of a sensorimotor system.

Another interesting model outline (Passingham and Toni 2001) defends that the prefrontal cortex (PFC) is critical for the integration between the streams. The motivation for this proposal is that the PFC lies at the top of the hierarchy of information processing, and receives information about cues, actions, and rewards. It is thus in the ideal position for integrating information about the external context, the action target and its outcome. The authors insist on the importance of the final goal on action selection. They point out that, when actions have to be mediated by explicit decisions, the contribution of cognitive information from the ventral stream and the PFC is required. Again, this model offers very plausible explanations of neuroscience findings, but does not suggest according to what mechanisms the integration between inferotemporal, posterior parietal and prefrontal cortices would actually be performed.

Cisek (2005) proposed a computational model for action selection in which the decision is taken through the interaction of posterior parietal, prefrontal and premotor cortex. The model is focused only on reaching movements, and although the implementation partially contradicts the indications given in Rizzolatti and Luppino (2001), some of the concepts introduced in this work could be adapted to action selection issues in grasping.

Recently, Oztop et al. (2006) have modeled the development of some neurons of the IPS (there is no explicit distinction between the tasks of CIP and AIP), matching depth maps directly to joint space. The model is built on a previous infant learning model (Oztop et al. 2004) which learns to match object affordances to hand postures. The main limits of these works are that affordances are very simplified, and actual visual features not considered, and the models are hence hardly transportable to a robotic setup. Nevertheless, the approach is interesting as the neural network implementation is able to learn visuomotor transformations and generalize to novel stimuli.

A similar visuomotor matching is developed in Uno et al. (1995), in which an artificial neural network learns the coupling between visual object data and hand postures. Again, both visual and motor representations are very basic, and although the model produces very relevant computational results, it seems to be not suitable for robotic implementation.

Overall, comparing the biologically-inspired robotic literature of Sect. 3.2 with the computational models regarding vision-based grasping, it looks as they work on different assumptions and with different goals. On the one hand, biological or neuroscientific inspiration in robotics is often too superficial and conditioned by pragmatic goals and technological constraints. On the other hand, computational models are usually focused on specific issues and simulate low-level processes that are hard to scale in order to produce complex behaviors, such as grasping. The model proposed in this work tries to achieve an intermediate and really interdisciplinary solution that—while maintaining biological plausibility, and the focus on neuroscience data, for the implementation of different visuomotor functions—provides the robot with the ability of performing efficient, flexible and reliable vision-based grasping.

4.2 Basic Modeling Concepts

In this section, fundamental computational principles, methodological decisions, restrictions and basic assumptions regarding implementation guidelines are provided.

4.2.1 Methodological Issues

A first important decision before beginning any implementation is to establish the level of detail and the techniques most suitable for the different stages of the model. It is critical to implement computational modules with the right level of abstraction. Models can be developed at virtually any level of detail, with different approaches and goals. For example, Bryson and Stein (2001) provide architectural guidelines for an implementation of brain areas as artificial agents. Such proposal has a strong point in the flexibility of the architecture, in which modules, like brain areas, can recruit one another according to actual requirements. An implementation of this kind would require nevertheless a complete homogeneity in the development of the different modules. We favor an approach focused on function rather than on structure.

In general, very detailed, low level implementations, as for example integrate and fire neuron models, are very good for reproducing neurophysiological data, but are not easily scalable for obtaining practical skills. On the other hand, very high level implementations rarely reproduce interesting effects. In fact, the functioning of the whole model should be robust, but also continuous, i.e., if noise is introduced, or links between areas modified, performance should degrade only gradually. Simple connectionist solutions may represent the required trade-off, as for example feed-forward neural networks, better with a biologically plausible architecture, such as radial basis functions (Pouget and Sejnowski 1997; Pouget and Snyder 2000). In any case, the suitability of the model to actual robot implementation has to be always taken into account. Modules can even have alternative implementations, as long as this does not affect the functioning of the whole system. As a criterion, an implementation that allows to emulate both high level and low level phenomena with good approximation should be favored. Chapter 5 provides results that exemplify this level of implementation.

As explained in the previous chapter, neuroscience research suggests that the bottom-up approach is suitable to model high-order cognitive processes. In fact, it seems that motor systems strongly contribute to the formation of processes traditionally considered to be "high level" or cognitive, such as action understanding, mental imagery of actions, perceiving and discriminating objects (Gallese et al. 1999). Mirror neurons are surely the best (and most famous) example of this. On the other hand, although bottom-up mechanisms account for much of the observable psychophysical data, the contribution of top-down stimuli can not be disregarded. Indeed, it has been argued (Goodale and Milner 2004; Freyet al. 2005; Arbib 2010) that classification, recognition and other more cognitive processes are used to trigger and accelerate the formation of brain-activity patterns relative to visuomotor tasks, including grasping (Sect. 2.5). The classification of an object within a certain class can facilitate or even partially bypass the process of grasp related feature extraction (Sugio et al. 2003a, b; Valyear and Culham 2010). In this book, although dorsal mechanisms are modeled with a higher level of detail, the contribution of the ventral stream is considered equally important and thus thoroughly taken into account.

4.2.2 Object, Hand, Task

In the grasping literature, three main factors affecting grasping actions are often cited: the object, the hand and the task (Mackenzie and Iberall 1994). The way in which each of these factors is taken into account in the proposed framework is described below.

4.2.2.1 Object

Two kinds of properties have to be considered for a potential target object. Spatial properties related to its current situation, such as distance and pose, can only be assessed through actual estimation. Implicit properties like its size, weight and consistence are instead obtained through the integration of on-line, instantaneous visual information with memory of previously acquired knowledge about the object. For grasping purposes, the on-line, dorsal elaboration consists mainly in a search for graspable zones, and the estimation of their size and position. Most often, graspable features can be found looking for approximately flat surfaces facing each others, or for cylindrical and spherical features.

It is commonly accepted that grip generation is a hierarchical process of variable complexity (Jeannerod 1997; Baud-Bovy and Soechting 2001; Gentilucci et al. 2003). The first processing step is the identification of an opposition axis on the target object (Iberallet al. 1986). It has been suggested that such opposition axis is effector independent, and it can be used for hand actions but also for other grasping means, such as biting (Castiello et al. 2000; Quinlan et al. 2005). Restraining to the hand case, opposing forces are applied by two *virtual fingers* (see box), selected accordingly to object geometry and task requirements.

A *virtual finger* is defined as an abstract functional representation of a collection of fingers and other hand parts applying an oppositional force (Arbib et al. 1985; Iberall 1987). Usually, the opposition is created between fingers and palm or between the thumb and one or more other fingers variably abducted. The thumb is likely the reference point for the movement in this last case (Galea et al. 2001; Ansuini et al. 2007a).

For identifying contact areas on the object surface, additional constraints have to be taken into account. Usually, an estimation of the object center of mass affects the grasp plan. Such estimation relies on data coming from the ventral pathway, as the expected object composition and density. Similarly, surface texture and thus the expected contact friction, which affect the required grasping force, are ventral stream information (McIntosh et al. 2004). Extraction and integration of different kinds of object properties is a central issue in the present model.

4.2.2.2 Hand

The geometry, kinematics and strength of the hand are of extreme importance in the task of grasp planning. Several taxonomies for classifying grasp types have been proposed, both in robotics and human studies (Cutkosky and Howe 1990).

Many different criteria can be used to define grasp categories, such as joint space, type of contact with the object, stability and so on. The only distinction that stands

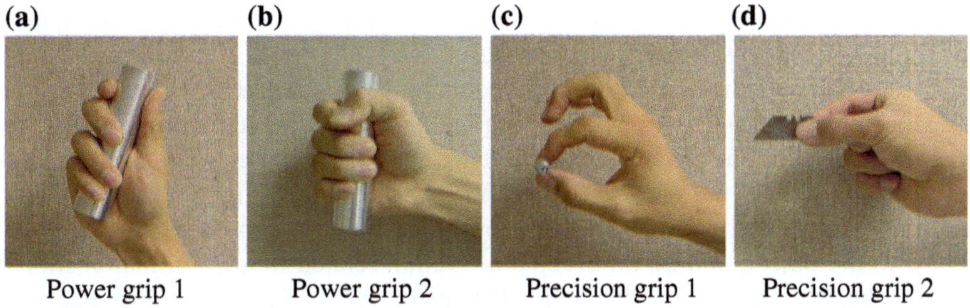

Fig. 4.5 Examples of power grips and precision grips. Images adapted from: http://www.bsu.edu/web/jkshim/handfinger/overview/overview.htm

up to all criteria is the separation between power and precision grips (Mackenzie and Iberall 1994). In power, or full-hand grips, the fingers create opposition with the palm of the hand. According to object shape and task, the thumb can either provide control (Fig. 4.5a) or reinforce the opposition space (Fig. 4.5b). The main opposition axis varies depending on object size and shape, but in general the grip includes support by the palm. The opposition axis in precision, or pinch grips, is created between the thumb and one or more fingers, especially the index finger. Again, control and force can be modulated: pure fingertip grips are extremely precise (Fig. 4.5c), while grips including finger pads are more stable (Fig. 4.5d).

Recent imaging studies show a differential activity in AIP between precision and power grips (Ehrsson et al. 2000; Begliomini et al. 2007; Cavina-Pratesi et al. 2007b), which seems to be not present in *pantomimed grasping* (see box).

> The expression *pantomimed grasping* is used to refer to simulated hand shaping, as for grasping, but without the presence of real objects. Despite the amount of research on simulated grasping and reaching movements, the mechanisms subserving pantomimed actions are still controversial (Westwood et al. 2000a; Tessari and Rumiati 2002; Milner et al. 2003; Makuuchi et al. 2012). Recent findings suggest that, at least for grasping, there is a visuomotor activation directly related to the presence of possible targets which is not observable in simulated actions without real goals (Takasawa et al. 2003; Cattaneo et al. 2005; Króliczak et al. 2007; Snow et al. 2011).

Two non-exclusive hypotheses, backed by other findings, can explain the different patterns of activity between real and pantomimed grasping. The first explanation is that differential activity between power and precision grips is strictly related to the visual analysis of the target object, even more than to the motor component. Precision grips require a more detailed visual analysis and hence the increased activity. Pantomimed grasping does not require visual analysis and no differential activation should be observed in this case. AIP fine responsiveness to orientation discrimination (Shikata et al. 2003) supports this hypothesis.

The second hypothesis regards the task component of the action. The fundamental influence of post-grasp tasks on hand shaping has been repeatedly proved (Ansuini et al. 2006, 2008). The basic distinction between power and precision grasping, fundamentally related to the action task, makes sense only if the action is performed with a real goal. In pantomimed grasping there is no actual goal provided by the interaction with the object, and no differential activation related to grip formation should then be observed. In fact, the distinction between power and precision grips has been regarded as mainly functional, and not separable from the initial goal of the action (Napier 1956). Power grips are performed, and the corresponding hand configuration designed for, exerting force on the target object, in order to achieve stability and security in the holding phase. Precision grips are instead suitable for accurate tactile response, and are executed pursuing the goals of dexterity and sensitivity (Cutkosky and Howe 1990).

The previous considerations indicate that it is very reductive to distinguish power and precision grips only on the basis of visual object features, unless in the case of extreme object size, like for very big or very small objects, that only afford large full-hand and small pinch grips respectively. The issue of grasp taxonomy related to object characteristics is further discussed in Sect. 7.1.

In any case, for most common tasks human grasping actions do not need an extremely high precision at first contact, as possible small misplacements are normally corrected on-line through tactile feedback. This simplifies grasp planning and leads to a reduced grip taxonomy, as similar grasping actions would be classified within the same grip class. In robotic grasping, small differences in finger positioning can lead to big changes in the quality of the grip (Morales et al. 2004), and thus precision is more critical, unless a haptic system to adjust the grip is available. Arguably, a real-world robotic grasping system can not prescind from good tactile skills, and furthermore, the generation of a big number of grips slightly different from each other is not only biologically implausible, but also extremely costly for an on-line grasping action. Therefore, a reduced grip taxonomy is the solution most plausible for neuroscience and most convenient for implementation.

The proposed model framework is based on the human hand, but restrictions posed by artificial robotic hands are taken into account where necessary. As explained in Sect. 7.1, the limits of a robotic hand strongly affect the number and quality of possible experiments.

4.2.2.3 Task

As already pointed out, the task is a critical, determinant factor in grasp planning (Ansuini 2008; Kawasaki et al. 2008; Ellison and Cowey 2009; Hoeren et al. 2013). There may be many appropriate ways of grasping an object, but if a given action has to be performed with it, it has to be grasped in a way that allows for the correct execution of that action. The concept of task is thus additional to visual grasp analysis, and surely biases the grip selection process. The handling of information about the task involves explicit knowledge regarding the object that has to be grasped and its potential use. Access to memories stored in the ventral pathway is thus required, but it is most likely the prefrontal cortex which deals with the complexity of integrating dorsal and ventral information while pursuing the execution of the required task (Binkofski et al. 1999; Lebedev and Wise 2002; Sereno et al. 2002).

In this work, the effect of task on grasp planning is simplified considering that a neutral task, such as "stably lift the object", has to be performed. Eventual requirements posed by subsequent actions are thus not taken into account.

4.2.3 Role of the Dorsal and Ventral Streams and Possible Interactions

The two streams perform different kinds of visual processing on primary visual input. The ventral stream performs a global, situation-invariant visual analysis and associates visual information to object identity, extracting perceptive information about the object nature. Ventral areas provide data about the object class, and its expected weight, roughness and compliance. Through recognition, previous grasping actions experienced on such object can also be recalled. The dorsal stream is instead oriented to a pragmatic spatial analysis of object local features. The products of its visual elaboration are precise information about position, shape and size of graspable object parts.

Using the extracted visual features, and possibly additional information derived from object identity, the dorsal pathway is responsible for generating possible affordances, or grip candidates, represented in a reference frame that can be directly used for action. Merging of circumstantial criteria with previous experience provides the mean through which the final action is selected and planned: on-line, perceptual data referred to geometry and position of the target object and perceptual information on relevant object properties have to be integrated in an efficient and robust way, in order to plan, execute and evaluate an appropriate grasping action. The complementary contributions of the two streams to this process are summarized in Table 4.1.

As mentioned in Sects. 2.4.3 and 2.5, many aspects affect the quantity and quality of tasks assigned to each stream in a given condition. In most cases the work partition between the streams is rather progressive, depending for example on task, action delay or on object familiarity (Sugio et al. 2003b; Himmelbach and Karnath 2005; Cohen et al. 2009; Tankus and Fried 2012; Wokke et al. 2014). An explanation for this last case is that contribution of the ventral stream on the formation of the grip is modulated by the confidence achieved by the LOC in the recognition of the target object. A higher confidence in object recognition reflects in a stronger influence of ventral stream data, such as knowledge of object weight and compliance. On the opposite, a more uncertain recognition leads to a more exploratory behavior, giving more importance to actual observation and dorsal analysis.

Table 4.1 Complementary tasks of the two streams

Ventral stream	Dorsal stream
Object recognition	Visuomotor control
Global, invariant analysis	Local, feature analysis
Object weight, roughness, compliance	Object local shape, size
Object meaning	Object location
Previous experiences	Actual working conditions
Scene-based frame of reference	Effector-based frame of reference
Long-term representation	On-line computation

The quality of the previous grasping experiences with an object has a similar effect. Proper action patterns can be recovered if recognition succeeds and consolidated experience of previous actions is available (Tucker and Ellis 2004). In these cases, there can be a strong bias toward learnt motor representations, that allows to partially bypass visual analysis. In any case, stored visuomotor patterns need to be refined with information gathered by the dorsal visual stream, which keeps its central role in detailed action planning and execution. For example, an object can be correctly recognized and immediately associated to a hand shape suitable for grasping it, but its location and orientation have still to be estimated employing on-line visual data.

It has been shown that humans are able to generate motor commands exactly suited to the weight of familiar objects (Eastough and Edwards 2007). Instead, for unfamiliar objects, subjects estimate their weight joining size information with assumptions on the possible object density, and perform an "average" grasp in order to reduce the risk of slipping or crushing the object (Gordon et al. 1993). In both cases there seem to be an important ventral contribution, either for recalling previous experience after object recognition, or for estimating object properties that can only be inferred if object composition is identified. The relation between the two streams in the proposed model builds on these concepts.

4.3 Model Framework

In the previous section, the scope and methodology of the proposal have been circumscribed. This section presents the framework of the whole model, which will be detailed in Chaps. 5 and 6. It is important to stress that, although the proposed approach is based on the neuroscience concepts introduced in Chap. 2, the following description of the system includes a number of hypotheses that should be confirmed by further neurophysiological and behavioral research.

Figure 4.6 shows the graphical schema of the whole model. The fundamental data flow is the following. After the extraction of basic visual information in V1/V2, higher level features are generated in V3/V3A and sent to the two streams. Along

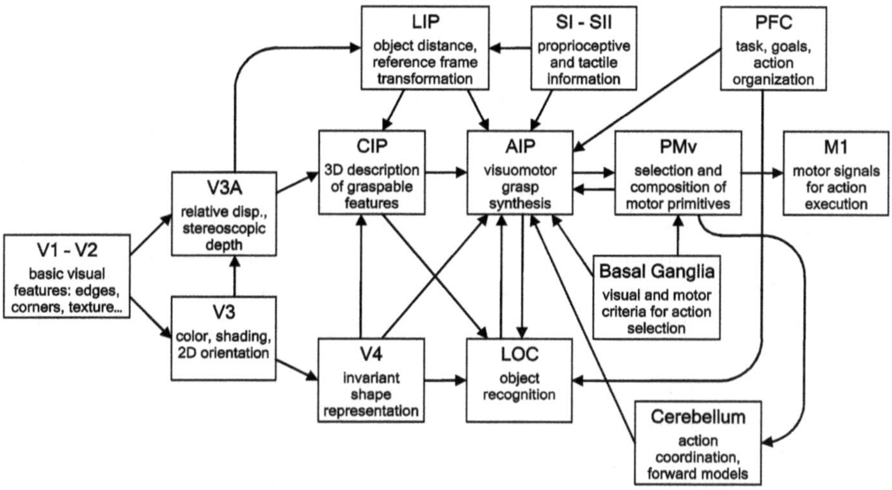

Fig. 4.6 Model Framework depicting all principal areas involved in the planning and execution of vision-based grasping actions

the ventral stream, an invariant representation of object shape is generated in order to perform a gradual recognition of the object. Area V4 is in charge of classifying the object as pertaining to a given class, and only later full recognition is performed in LOC. In the dorsal stream, area CIP integrates stereoptic and perspective data in order to detect pose and proportion of the target object, using also information regarding object classification. Area LIP estimates object location integrating retinal data with proprioceptive information about eye position. Both CIP and LIP project to AIP, which transforms object visual data in hand configurations suitable for grasping. A grasping plan is devised by AIP in collaboration with PMv, considering also the information on object identity coming from the ventral stream, and task requirements given by the prefrontal cortex, PFC. Dorsal areas are supported by proprioceptive information coming from somatosensory areas SI/SII, and action selection in AIP and PMv is modulated by the basal ganglia (BG). The signals for action execution are sent to the motor cortex M1, and an AIP-PMv-Cerebellum loop is in charge of monitoring action execution in accordance to the plan.

Blocks in the diagram correspond to brain areas, according to the findings described in Chap. 2, but also to implementable functional modules. In the following sections, basic concepts regarding the functioning, the role and the interactions of the brain areas which appear in the model are given. The basic interest is directed toward the integration of different types of information as it is thought to happen through the connections between the dorsal and the ventral streams. Transfer functions and implementation details will be provided in the next two chapters.

4.3.1 Processing of Basic Visual Information

The first step is the extraction of fundamental visual data regarding the target object. Starting from visual acquisition, an attentional mechanism is needed to focus on the object to grasp, for isolating it from the background and possible other objects. Once the object is unambiguously identified, visual elaboration can begin.

Visual areas V1 and V2 receive images and provide as output basic features, such as edges, corners, and absolute disparities. These features are used by downstream areas to build more complex ones. The most advanced visual representation common to both streams is a basic binocular description of the target object, composed for both eyes of its contour as a 2D silhouette and the retinal position of salient features, such as sharp corners. After this stage, the visual analysis is performed in parallel concurrent ways by the two pathways.

The ventral stream performs a gradual classification and identification of objects, probably through the integration of volumetric descriptions with 2D ones, as described in Sect. 2.4.1. On the other hand, the search of the dorsal stream for places on which to put the fingers is better done on descriptions of objects represented by 2D surfaces disposed in the 3D space (see Sect. 2.3.1). Color information, processed by area V3, can be used by the ventral stream to recognize objects more easily, and by the dorsal stream to track objects, but also to extract surface properties through shading and textures.

4.3.2 Extraction of Visual Features Suitable for Grasping

The description of visual object features relevant for grasping purposes is the next processing stage. The posterior parietal cortex, in charge of this task, does not construct any model or global representation of the object, but rather extract properties of visual features that are suitable for a potential action. In order to elaborate a proper action on an external target, two main inputs are required, the object shape and pose and its location with respect to the eyes and thus to the hand. These inputs are obtained by integrating retinal information regarding the object with proprioceptive data referred to eyes, head and hand. All this information is managed contextually by the dorsal stream.

4.3.2.1 Estimation of Geometry and Location of the Target Object

Visual area V3A is retinotopically organized, so that topological distribution of the observed scene, and the relative position of object features are directly represented. Relative disparities, fundamental for pose estimation, are also calculated in V3A. Basic representations of visible surfaces coming from V3A are the input of intraparietal areas. In humans, the intraparietal sulcus is where visual input begins to be coded in a head-based spatial representation. The lateral intraparietal area LIP seems to be responsible for remapping retinal visual data in eye-centered coordinates (Sect. 2.3.4), and for estimating object location and distance. Cues to distance estimation are retinal data, accommodation and vergence, this last being probably more influent in the dorsal stream, especially for grasping distances (Tresilian et al. 1999; Goodale and Milner 2004). A model of distance estimation from proprioceptive vergence data is proposed in Chap. 5.

Area CIP elaborates the visual input it receives, mainly from V3A, in order to produce features that are possibly suitable for grasping purposes. This area is in charge of performing complex orientation discrimination of 3D object features, and codes visual information in a format suitable to be used by AIP for generating appropriate hand configurations and grasping movements. The outstanding human grasping abilities in open-loop conditions suggest that visual analysis of object features is performed very accurately. Increased reliability in estimation can be achieved using a multiple cue approach, in which concurrent cues are optimally integrated to produce a better estimate. Details on the integration of stereoptic and perspective cues for pose estimation in CIP are given in Chap. 5.

Features extracted by CIP are coded using the response properties of SOS and AOS neurons, which are selective for the orientation and relative proportion of object parts (Sect. 2.3.1). Some responsiveness to actual object size has also been observed, suggesting that at least a part of AOS and SOS neurons code for absolute object dimensions. Although an exact object size representation is most likely maintained in AIP, approximate size data could be used in CIP to rule out ungraspable features, as discussed in Sect. 6.2.1. Summarizing, the output of CIP, conveyed by AOS and SOS neurons, are the orientation and relative dimensions of features, filtered according to their suitability for grasping actions. This is the major input to AIP, which will transform it in suitable hand shapes. Reference frame transformation from head-based to hand-based is probably executed in AIP as well.

4.3.2.2 First Interaction Between the Streams

Visual processing in the ventral stream is based on the production of increasingly invariant representations aimed at object recognition. During grasping actions, ventral visual areas are in charge of identifying the object, and facilitating access to memorized properties which can be useful for the oncoming action. Region V4 codes at the same time shape, color and texture of features, which are then composed in the LOC to form more complex representations recognizable as objects. Output from area V3 is thus used by V4 to build a viewpoint invariant simple coding of the object, that can be used to classify it as belonging to one of a number of known object classes. Basic computational representations for this purpose are for example chain codes or 2D shape indexes, as explained in Sect. 5.2.3.

Information on the basic shape of the object is probably forwarded to the dorsal stream, to CIP or AIP or both, to facilitate the feature extraction process. For example, if the object is recognized as roughly box-like, it can be assumed that its edges are parallel. Such assumption would facilitate the process of size and pose estimation, because reliable perspective estimation can be used in this case in addition to stereopsis.

Downstream from V4, the LOC compares spatial and color data with stored information about previously observed objects, to finally recognize the target as a single, already encountered object. Object identification is thus performed in a hierarchical fashion (see Sect. 5.1.3), where the target is first classified into a given class and, only later, exactly identified as a concrete object. In each of these steps, recognition is not a true/false decision, but rather a probabilistic process, in which an object is classified or identified only up to a given confidence level. Thus, confidence values should be provided by the classification and identification procedures. In this way, ventral information can be given more or less credit. If recognition confidence is high, visual analysis can be simplified, as most required information regarding the target object is already available in memory. If recognition is instead considered unreliable, more importance is given to the on-line visual analysis performed by the dorsal stream.

Final output of the object recognition process are its identity and composition, which in turn allows to estimate its weight distribution and the roughness of its surface, that are valuable information at the moment of planning the action. Moreover, besides the recovery of memorized object properties, recognition allows to access stored knowledge regarding previous grasping experiences. Old actions on that object can be recalled and used to bias grasp selection, giving preference to learnt hand configurations which ended in successful action executions. Similarly to the classification confidence, the number and outcome of previous encounters with the same object will determine the reliability of the stored information.

4.3.3 Transformation of Visual Features into Hand Shapes

Although some AIP neurons code for specific spatial aspects, similarly to CIP's, orientation and shape are often represented as a combined 3D feature in AIP. Researchers concord in the interpretation that such combined coding is suitable for representing graspable features at a visuomotor level (Castiello 2005; Sakata et al. 2005). As largely justified in Chap. 2, AIP is the area of the primate brain which performs the visuomotor transformations necessary for grasping actions. Less clear is how much of the information that AIP requires in order to plan an action is provided by CIP, and how much needs to be complemented by other areas. Apart for the known circuit linking AIP to the ventral premotor cortex, several other brain regions show major projections to the anterior intraparietal area. AIP surely receives information from other areas of the intraparietal sulcus, from the ventral stream, the inferior parietal lobe, the basal ganglia, the cerebellum and the prefrontal cortex. The information coming from LIP and the inferior parietal lobe is of visual nature, and the data proceeding from the ventral stream are about characteristics of the object. Connections with PMv and cerebellum are more related with motor aspects of the action, and the basal ganglia probably participates in action selection. The prefrontal cortex organizes action plans according to the context, and sets goals and tasks. The relevance and effect of such inputs to data processing in AIP will be discussed in the next two chapters. In the next section, some visual and motor aspects used in grasp synthesis and assessment are described.

4.3.3.1 Grasping Constraints and Grasp Quality

Provided that AIP is mostly concerned with precision grasping (Sect. 4.2.2.2), it makes sense that its major source of information in order to generate suitable hand shapes for the selected target feature is the precise responsiveness of SOS and AOS neurons of CIP. Features can be classified, although not exclusively, as flat or elongated ones according to such responsiveness. Exact size estimation, likely provided by LIP, and CIP output are used to establish a hand shape suitable for the selected feature. Integration of these data with proprioceptive input regarding the relative location of object and hand, and the current position and state of hand and arm, allows to perform the transformations necessary to recode visual data in an actor-based reference frame.

For what concerns the selection of contact places between hand and object, humans usually try to grasp objects so that the opposition axis passes close to the estimated center of mass, and perpendicular to the main inertia axes (Bingham and Muchisky 1993a; Wagman and Carello 2003; Freyet al. 2005). More exactly, subjects tend to align the grip axis with the minor axis of the object (Cuijpers et al. 2004). For flat objects, having predominant SOS activations, the preferred opposition axis is parallel to the minor size (Fig. 4.7a). For elongated objects, corresponding to high AOS activations, the grasping axis is usually perpendicular to the major size (Fig. 4.7b). In this way, misalignment errors have less effect on the possible consequent torques as compared with grasping aligned to the major axes.

Another subgoal of grasp selection is to place the fingers in safe positions on the object. In many cases this is not an issue, especially for large and regular objects.

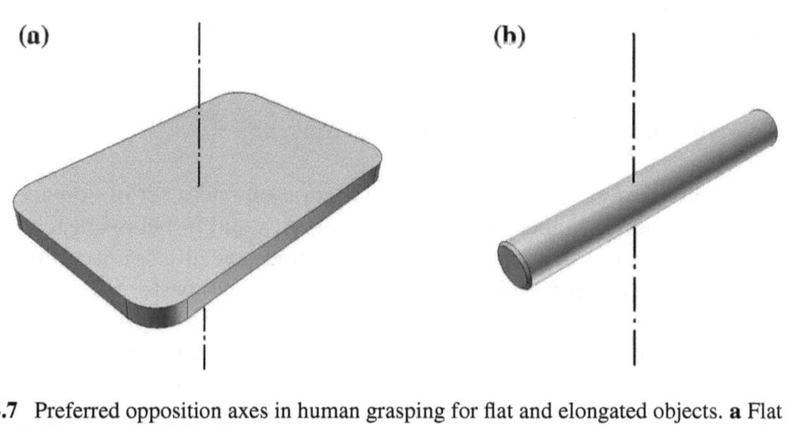

Fig. 4.7 Preferred opposition axes in human grasping for flat and elongated objects. **a** Flat (SOS) objects. **b** Elongated (AOS) objects

Moreover, grasping close to the center of mass very often ensures that contacts are far from the edges as well. If the object is small though, the risk of placing a finger on an edge, and thus reducing the grasp reliability, is real. This same example also reveals the preference toward concave compared to convex features for improving grasping stability (Jenmalm et al. 2000). Curvature is very likely included in information forwarded from CIP to AIP, which can use it during the selection of contact areas. For complex, asymmetric and irregular objects, all these aspects have to be taken thoroughly into account.

Regarding arm and hand posture, subjects tend to pursue a sort of reaching "comfort", adapting to a natural grasp axis which minimizes the rotation of hand and arm with respect to the starting position (Cuijpers et al. 2004). Such preferred approaching direction is respected as much as the object allows for it. In the case of spherical objects, or of cylindrical objects placed vertically, the grasp axis is found to be always consistent with this preferred reaching orientation (Paulignan et al. 1997). If objects are not symmetrical, and possess for example a clear opposition axis, then a modulation between effect of shape and position is observed (Ledermana nd Wing 2003). Both visual and motor criteria hence constrain the selection of the grip shape and the contact with the object. It appears that, if AOS activation is prevailing, the motor criterion of comfortable reaching is probably dominant. Instead, in case of stronger SOS response, the reaching direction will have to adapt to the opposition axis afforded by the object. Visual and motor criteria similar to those described in this section can also be exploited to select between different candidate grasping features pertaining to the same object. There is most probably a contribution of the basal ganglia to this action selection process, namely of its output nucleus *substantia nigra pars reticulata* (Clower et al. 2005). More details on the criteria for action selection and their use are provided in Sect. 6.2.2.

Together with the selected grip, an initial grasping force is also provided in the grasp planning process. In effect, it seems that AIP is also responsible for determining the initial force of grips (Ehrsson et al. 2003; Goodale and Milner 2004), as this is strictly related to the geometry of the grasping configuration and the nature of the target object. Force is later adapted to the task requirements and to object characteristics after contact has been established and tactile feedback is available.

4.3.3.2 Second Interaction Between the Streams

The general problem of selecting a hand configuration and a set of contact points for grasping geometric shapes is common in robotics and in neuroscience experiments. Nevertheless, in real grasping this rather abstract selection is not common, and cognitive reasoning about the object quality and the task to perform usually comes before grasp decisions. Hence, in everyday life actions are strongly affected, if not dominated, by ventral stream processing, and the influence of LOC on AIP is very strong. Still, for most objects, only excluding extremely symmetrical ones, a visual analysis for grasping is necessary in order to detect their pose, and if high precision is required in finger placing, such analysis has to be quite accurate. The continuous interaction between the perceptual and visuomotor systems is thus the most likely normal way of functioning for vision based grasping (Goodale et al. 2004; Wokke et al. 2014; Zachariou et al. 2014).

The visuomotor factors affecting grasp planning, described in Sect. 4.3.3.1, are constantly biased by experience and conceptual knowledge provided by the ventral stream. As explained in Sect. 4.3.2.2, once the object has been recognized, information concerning its properties and memories of previous grasping experiences can be recalled. Confidence in the recognition and quality of memories modulate the effect that ventral stream data has on the grasping action. If the recognition confidence is high, more importance can be given to some visuomotor criteria with respect to others. An object recognized as heavy will elicit a grasping action in which closeness to the center of mass is very important, even though the reaching movement is less comfortable. Similarly, initial grasping force will be high in order to adapt to the characteristics of the target. In the case of uncertain object recognition, or encounters with unknown objects, a more conservative approach is probably used. The initial force will probably be lighter, in order to avoid crushing the object, and the action will be slower. After contact, tactile feedback complements the missing information and allows to properly carry out the planned action.

The same happens with knowledge regarding the roughness of the object, which drives the contact friction, and hence again the force required to grasp it. Many other ventral factors can bias the importance of the different visuomotor criteria. For example, object fragility or the presence of potentially dangerous features such as sharp edges, affect finger placement and contact force.

4.3.4 Grasp Execution

The final grasping plan is likely devised by AIP in collaboration with PMv and with the ventral stream, which has to build the memory of previous encounters with the target object for future reference. The AIP → LOC projection in Fig. 4.6 symbolizes this process. A mechanism like the one introduced in Sect. 3.3 could stand behind the symbolic representation of grasping actions required to store their memory both at a sensorimotor and at a cognitive level.

As mentioned in Sect. 2.3.2, AIP plays an active role in action execution. The action progress can be followed visually and, after contact, also haptically. AIP is most probably in charge of both monitoring modalities, and the same motor program is probably shared by AIP and PMv, maybe as a corollary discharge sent from the latter to the former. PMv is in charge of sending motor signals to the primary motor area and keeping it updated with the most recent motor plans. Meanwhile, AIP performs a comparison between expected sensory states and real visual and somatosensory stimuli, monitoring the progress of the ongoing action with respect to the plans (Ogawa et al. 2007). Such monitoring can be performed using forward models, likely maintained in the cerebellum. The same signals that PMv sends to M1 are sent also to the cerebellum, that is able to build a prediction of the sensory feedback in each instant of the execution. The cerebellum is believed to be the place in which forward models are computed, and evidence for projections from the cerebellum to AIP supports this schema (Clower et al. 2005). In case of sensory discrepancies or changes in the environment, suitable motor commands for correction are calculated, and the motor plan is updated. The ventral premotor cortex of the human brain has often been considered to make large use of inverse model for motor programming purposes (Kawato 1999; Miall 2003) and is probably able to change dynamically its motor plans according to AIP requirements.

In case of imprecise grasping, tactile adjustment after contact can also be based upon a closed-loop finger position updating, in which a goal state is calculated through forward models, and compared to the actual sensory feedback. Inverse models are used to devise the motor plan most appropriate for reducing the discordance between expected and real stimuli, and thus achieve the final stable state. A simple example of this mechanism is provided in Sect. 6.2.4.2. Once grasp is secured, the object can be lifted and any post-grasping task can be executed.

The quality of the contacts between fingers and objects can be immediately assessed, as will be shown in Sect. 6.2.4.2. This assessment is an important source of information for future actions. If the object was slightly displaced with respect to the expected position, it means that pose estimation was probably affected by some imprecision. Contextual aspects of the action will be associated with such imprecision, and after a few misplaced executions a pattern of error causes can be extracted from experience. In this way, the actual outcome of action execution is used to update the memory of accumulated grasping experience for future reference. The reward system, that establishes the most important goals to pursue in each condition, is also updated at this stage.

4.4 Conclusions

Computational models of the human visual system are largely available, especially for the first stages of visual processing, before the splitting of the two streams. At the same time, research on object recognition keeps involving a large part of the computer vision community. Nevertheless, few resources have been dedicated to the exploration of the mechanisms underlying the functioning of the action-related visual cortex, and the integration between the contributions of the two visual pathways is nearly unexplored at the computational level and even more in robotics. Thanks to recent neuroscience findings, the outline of a model of the brain mechanisms upon which vision-based grasp planning relies could be drawn in this chapter.

With respect to the available models, the proposed framework has been conceived to be applied on a robotic setup, and the analysis of the functions of each brain area has been performed taking into account not only biological plausibility, but also practical issues related to engineering constraints. In the next two chapters, implementation details and experimental results will be provided.

References

Ansuini C (2008) Reaching beyond grasp. Ph.D. thesis, Università degli Studi di Padova
Ansuini C, Santello M, Massaccesi S, Castiello U (2006) Effects of end-goal on hand shaping. J Neurophysiol 95(4):2456–2465. doi:10.1152/jn.01107.2005
Ansuini C, Giosa L, Turella L, Alto G, Castiello U (2008) An object for an action, the same object for other actions: effects on hand shaping. Exp Brain Res 185(1):111–119. doi:10.1007/s00221-007-1136-4
Ansuini C, Santello M, Tubaldi F, Massaccesi S, Castiello U (2007a) Control of hand shaping in response to object shape perturbation. Exp Brain Res 180(1):85–96, 006–0840-9
Arbib MA, Iberall T, Lyons D (1985) Coordinated control programs for control of the hands. In: Goodwin AW, Darian-Smith I (eds) Hand function and the neocortex. Experimental brain research supplemental, vol 10. Springer, Berlin, pp 111–129
Arbib MA (2010) Mirror system activity for action and language is embedded in the integration of dorsal and ventral pathways. Brain Lang 112(1):12–24. doi:10.1016/j.bandl.2009.10.001
Baud-Bovy G, Soechting JF (2001) Two virtual fingers in the control of the tripod grasp. J Neurophysiol 86(2):604–615
Begliomini C, Wall MB, Smith AT, Castiello U (2007) Differential cortical activity for precision and whole-hand visually guided grasping in humans. Eur J Neurosci 25(4):1245–1252. doi:10.1111/j.1460-9568.2007.05365.x
Bingham GP, Muchisky MM (1993a) Center of mass perception and inertial frames of reference. Percept Psychophysi 54(5):617–632
Binkofski F, Buccino G, Posse S, Seitz RJ, Rizzolatti G, Freund H (1999) A fronto-parietal circuit for object manipulation in man: evidence from an fMRI-study. Eur J Neurosci 11(9):3276–3286
Bryson J, Stein LA (2001) Modularity and specialized learning: mapping between agent architectures and brain organization. Emergent Neural Comput Architectures Based Neurosci LNCS 2036:98–113
Cadieu C, Kouh M, Pasupathy A, Connor CE, Riesenhuber M, Poggio T (2007) A model of V4 shape selectivity and invariance. J Neurophysiol 98(3):1733–1750. doi:10.1152/jn.01265.2006
Castiello U, Bennetta KM, Egan GF, Tochon-Danguy HJ, Kritikos A, Dunai J (2000) Human inferior parietal cortex programs the action class of grasping. Cogn Syst Res 1(2):89–97. doi:10.1016/S1389-0417(99)00011-X
Castiello U (2005) The neuroscience of grasping. Nat Rev Neurosci 6(9):726–736. doi:10.1038/nrn1744
Cattaneo L, Voss M, Brochier T, Prabhu G, Wolpert DM, Lemon RN (2005) A cortico-cortical mechanism mediating object-driven grasp in humans. Proc Nat Acad Sci USA 102(3):898–903. doi:10.1073/pnas.0409182102
Cavina-Pratesi C, Monaco S, McAdam T, Milner D, Schenk T, Culham JC (2007b) Which aspects of hand-preshaping does human AIP compute during visually guided actions? Evidence from event-related fMRI. In: Annual meeting of the society for neuroscience
Chinellato E, Fisher RB, Morales A, del Pobil AP (2003a) Ranking planar grasp configurations for a three-finger hand. In: IEEE International conference on robotics and automation, Taipei, Taiwan
Cisek P (2005) A computational model of reach decisions in the primate cerebral cortex. In: Modeling natural action selection
Clower DM, Dum RP, Strick PL (2005) Basal ganglia and cerebellar inputs to 'AIP'. Cereb Cortex 15(7):913–920. doi:10.1093/cercor/bhh190
Cohen NR, Cross ES, Tunik E, Grafton ST, Culham JC (2009) Ventral and dorsal stream contributions to the online control of immediate and delayed grasping: a tms approach. Neuropsychologia 47(6):1553–1562. doi:10.1016/j.neuropsychologia.2008.12.034
Cuijpers RH, Smeets JBJ, Brenner E (2004) On the relation between object shape and grasping kinematics. J Neurophysiol 91(6):2598–2606. doi:10.1152/jn.00644.2003

Cutkosky M, Howe R (1990) Human grasp choice and robotic grasp analysis. In: Venkataraman S, Iberall T (eds) Dextrous robot hands, Springer, Berlin, Chap. 1, pp 5–31

Eastough D, Edwards MG (2007) Movement kinematics in prehension are affected by grasping objects of different mass. Exp Brain Res 176(1):193–198. doi:10.1007/s00221-006-0749-3

Ehrsson HH, Fagergren A, Jonsson T, Westling G, Johansson RS, Forssberg H (2000) Cortical activity in precision- versus power-grip tasks: an fMRI study. J Neurophysiol 83(1):528–536

Ehrsson HH, Fagergren A, Johansson RS, Forssberg H (2003) Evidence for the involvement of the posterior parietal cortex in coordination of fingertip forces for grasp stability in manipulation. J Neurophysiol 90(5):2978–2986. doi:10.1152/jn.00958.2002

Ellison A, Cowey A (2009) Differential and co-involvement of areas of the temporal and parietal streams in visual tasks. Neuropsychologia 47(6):1609–1614. doi:10.1016/j.neuropsychologia.2008.12.013

Fagg AH, Arbib MA (1998) Modeling parietal-premotor interactions in primate control of grasping. Neural Networks 11(7–8):1277–1303

Frey SH, Vinton D, Norlund R, Grafton ST (2005) Cortical topography of human anterior intraparietal cortex active during visually guided grasping. Cogn Brain Res 23(2–3):397–405. doi:10.1016/j.cogbrainres.2004.11.010

Fukuda H, Fukumura N, Katayama M, Uno Y (2000) Relation between object recognition and formation of hand shape: a computational approach to human grasping movements. Syst Comput Jpn 31(12):11–22

Fukui T, Takemura N, Inui T (2006) Visuomotor transformation process in goal-directed prehension: utilization of online vision during preshaping phase of grasping. Jpn Psychol Res 48(3):188–203. doi:10.1111/j.1468-5884.2006.00318.x, http://www.blackwell-synergy.com/doi/pdf/10.1111/j.1468-5884.2006.00318.x

Galea MP, Castiello U, Dalwood N (2001) Thumb invariance during prehension movement: effects of object orientation. Neuroreport 12(10):2185–2187

Gallese V, Craighero L, Fadiga L, Fogassi L (1999) Perception through action. Psyche 5(21):1

Gentilucci M, Caselli L, Secchi C (2003) Finger control in the tripod grasp. Exp Brain Res 149(3):351–360. doi:10.1007/s00221-002-1359-3

Goodale MA, Milner AD (2004) Sight unseen. Oxford University Press, Oxford

Goodale MA, Westwood DA, Milner AD (2004) Two distinct modes of control for object-directed action. Prog Brain Res 144:131–144

Gordon AM, Westling G, Cole KJ, Johansson RS (1993) Memory representations underlying motor commands used during manipulation of common and novel objects. J Neurophysiol 69(6):1789–1796

Himmelbach M, Karnath HO (2005) Dorsal and ventral stream interaction: contributions from optic ataxia. J Cogn Neurosci 17(4):632–640. doi:10.1162/0898929053467514

Hoeren M, Kaller CP, Glauche V, Vry MS, Rijntjes M, Hamzei F, Weiller C (2013) Action semantics and movement characteristics engage distinct processing streams during the observation of tool use. Exp Brain Res 229(2):243–260. doi:10.1007/s00221-013-3610-5

Hubel D, Wiesel T (1962) Receptive fields, binocular interaction and functional architecture in the cat's visual cortex. J Physiol London 160(1):106–154

Iberall T (1987) The nature of human prehension: three dextrous hands in one. In: IEEE international conference on robotics and automation, pp 396–401

Iberall T, Bingham G, Arbib MA (1986) Opposition space as a structuring concept for the analysis of skilled hand movements. In: Heuer H, Fromm C (eds) Generation and modulation of action patterns. Springer, Berlin, pp 158–173

Jeannerod M (1997) The cognitive neuroscience of action. Blackwell, Oxford

Jenmalm P, Dahlstedt S, Johansson RS (2000) Visual and tactile information about object-curvature control fingertip forces and grasp kinematics in human dexterous manipulation. J Neurophysiol 84(6):2984–2997

Kawasaki M, Watanabe M, Okuda J, Sakagami M, Aihara K (2008) Human posterior parietal cortex maintains color, shape and motion in visual short-term memory. Brain Res 1213:91–97. doi:10.1016/j.brainres.2008.03.037

Kawato M (1999) Internal models for motor control and trajectory planning. Curr Opin Neurobiol 9:718–727

Khosla D, Chen Y, Kim K (2014) A neuromorphic system for video object recognition. Front Comput Neurosci 8:147. doi:10.3389/fncom.2014.00147

Króliczak G, Cavina-Pratesi C, Goodman DA, Culham JC (2007) What does the brain do when you fake it? an fMRI study of pantomimed and real grasping. J Neurophysiol 97(3):2410–2422. doi:10.1152/jn.00778.2006

Lebedev MA, Wise SP (2002) Insights into seeing and grasping: Distinguishing the neural correlates of perception and action. Behav Cognitive Neurosci Rev 1(2):108–129. doi:10.1177/1534582302001002002

Lederman SJ, Wing AM (2003) Perceptual judgement, grasp point selection and object symmetry. Exp Brain Res 152(2):156–165. doi:10.1007/s00221-003-1522-5

Lourens T, Barakova EI (2007) Orientation contrast sensitive cells in primate V1 a computational model. Nat Comput 6(3):241–252

Mackenzie C, Iberall T (1994) The grasping hand. North Holland, New York

Makuuchi M, Someya Y, Ogawa S, Takayama Y (2012) Hand shape selection in pantomimed grasping: interaction between the dorsal and the ventral visual streams and convergence on the ventral premotor area. Hum Brain Mapp 33(8):1821–1833. doi:10.1002/hbm.21323

McIntosh RD, Dijkerman HC, Mon-Williams M, Milner AD (2004) Grasping what is graspable: evidence from visual form agnosia. Cortex 40(4–5):695–702

McLaughlin D, Shapley R, Shelley M (2003) Large-scale modeling of the primary visual cortex: influence of cortical architecture upon neuronal response. J Physiol Paris 97(2–3):237–252. doi:10.1016/j.jphysparis.2003.09.019

Miall R (2003) Connecting mirror neurons and forward models. Neuroreport 14(16):1–3

Milner AD, Dijkerman HC, McIntosh RD, Rossetti Y, Pisella L (2003) Delayed reaching and grasping in patients with optic ataxia. Prog Brain Res 142:225–242

Morales A, Sanz PJ, del Pobil AP, Fagg AH (2006) Vision-based three-finger grasp synthesis constrained by hand geometry. Rob Auton Syst 54(6):496–512

Morales A, Chinellato E, Fagg AH, del Pobil AP (2004) Using experience for assessing grasp reliability. Int J Humanoid Rob 1(4):671–691

Napier JR (1956) The prehensile movements of the human hand. J Bone Joint Surg (Br) 38-B(4):902–913

Ogawa K, Inui T, Sugio T, (2007) Neural correlates of state estimation in visually guided movements: an event-related fMRI study. Cortex 43(3):289–300

O'Reilly RC, Munakata Y (2000) Computational explorations in cognitive neuroscience—understanding the mind by simulating the brain. MIT Press, Cambridge

Oztop E, Bradley NS, Arbib MA (2004) Infant grasp learning: a computational model. Exp Brain Res 158(4):480–503. doi:10.1007/s00221-004-1914-1

Oztop E, Arbib MA (2002) Schema design and implementation of the grasp-related mirror neuron system. Biol Cybern 87(2):116–140. doi:10.1007/s00422-002-0318-1

Oztop E, Imamizu H, Cheng G, Kawato M (2006) A computational model of anterior intraparietal (AIP) neurons. Neurocomputing 69(10–12):1354–1361

Passingham RE, Toni I (2001) Contrasting the dorsal and ventral visual systems: guidance of movement versus decision making. Neuroimage 14(1 Pt 2):S125–S131. doi:10.1006/nimg.2001.0836

Paulignan Y, Frak VG, Toni I, Jeannerod M (1997) Influence of object position and size on human prehension movements. Exp Brain Res 114(2):226–234

Pouget S, Sejnowski A (1997) Spatial transformations in the parietal cortex using basis functions. J Cogn Neurosci 9(2):222–237

Pouget A, Snyder LH (2000) Computational approaches to sensorimotor transformations. Nat Neurosci 3(Suppl):1192–1198. doi:10.1038/81469

Quinlan DJ, Goodale MA, Culham JC (2005) Don't bite the hand that feeds you: a comparison of mouth and hand kinematics. J Vis 5(8):382

Riesenhuber M, Poggio T (2000) Models of object recognition. Nat Neurosci 3:1199–1204. doi:10.1038/81479

Rizzolatti G, Luppino G (2001) The cortical motor system. Neuron 31(6):889–901

Rolls E, Deco G (2002) Computational neuroscience of vision. Oxford University Press, Oxford, UK

Rolls ET, Webb TJ (2014) Finding and recognizing objects in natural scenes: complementary computations in the dorsal and ventral visual systems. Front Comput Neurosci 8:85. doi:10.3389/fncom.2014.00085

Sabatini SP, Solari F, Andreani G, Bartolozzi C, Bisio GM (2001) A hierarchical model of complex cells in visual cortex for the binocular perception of motion-in-depth. Adv Neural In Process Syst 1271–1278

Sakata H, Tsutsui KI, Taira M (2005) Toward an understanding of the neural processing for 3D shape perception. Neuropsychologia 43(2):151–161. doi:10.1016/j.neuropsychologia.2004.11.003

Sereno ME, Trinath T, Augath M, Logothetis NK (2002) Three-dimensional shape representation in monkey cortex. Neuron 33(4):635–652

Shikata E, Hamzei F, Glauche V, Koch M, Weiller C, Binkofski F, Büchel C (2003) Functional properties and interaction of the anterior and posterior intraparietal areas in humans. Eur J Neurosci 17(5):1105–1110

Snow JC, Pettypiece CE, McAdam TD, McLean AD, Stroman PW, Goodale MA, Culham JC (2011) Bringing the real world into the fmri scanner: repetition effects for pictures versus real objects. Sci Rep 1:130. doi:10.1038/srep00130

Sugio T, Ogawa K, Inui T (2003a) Neural correlates of semantic effects on grasping familiar objects. Neuroreport 14(18):2297–2301. doi:10.1097/01.wnr.0000092474.09492.3a

Sugio T, Ogawa K, Inui T (2003b) Multiple action representations of familiar objects with handles: an fMRI study. In: European conference on visual perception

Takasawa M, Oku N, Osaki Y, Kinoshita H, Imaizumi M, Yoshikawa T, Kimura Y, Kajimoto K, Sasagaki M, Kitagawa K, Hori M, Hatazawa J (2003) Cerebral and cerebellar activation in power and precision grip movements: an $H_2\,^{15}O$ positron emission tomography study. J Cereb Blood Flow Metab 23(11):1378–1382, doi:DOIurl10.1097/01.WCB.0000091258.83091.C2

Tankus A, Fried I (2012) Visuomotor coordination and motor representation by human temporal lobe neurons. J Cogn Neurosci 24(3):600–610

Tessari A, Rumiati RI (2002) Motor distal component and pragmatic representation of objects. Cogn Brain Res 14(2):218–227

Tresilian JR, Mon-Williams M, Kelly BM (1999) Increasing confidence in vergence as a cue to distance. Proc R Soc B: Biol Sci 266(1414):39–44

Tucker M, Ellis R (2004) Action priming by briefly presented objects. ACTA Psychol (Amsterdam) 116(2):185–203. doi:10.1016/j.actpsy.2004.01.004

Ullman S (1996) High-level vision. Object recognition and visual cognition. MIT Press, Cambridge

Uno Y, Fukumura N, Suzuki R, Kawato M (1995) A computational model for recognizing objects and planning hand shapes in grasping movements. Neural Networks 8(6):839–851. doi:10.1016/0893-6080(95)00002-H

Valyear KF, Culham JC (2010) Observing learned object-specific functional grasps preferentially activates the ventral stream. J Cogn Neurosci 22(5):970–984. doi:10.1162/jocn.2009.21256

Wagman JB, Carello C (2003) Haptically creating affordances: the user-tool interface. J Exp Psychol: Appl 9(3):175–186. doi:10.1037/1076-898X.9.3.175

Westwood DA, Chapman CD, Roy EA (2000a) Pantomimed actions may be controlled by the ventral visual stream. Exp Brain Res 130(4):545–548

Wokke ME, Scholte HS, Lamme VAF (2014) Opposing dorsal/ventral stream dynamics during figure-ground segregation. J Cogn Neurosci 26(2):365–379

Yousefi B, Loo CK (2014a) Comparative study on interaction of form and motion processing streams by applying two different classifiers in mechanism for recognition of biological movement. Sci World J 723:213. doi:10.1155/2014/723213

Yousefi B, Loo CK (2014b) Development of biological movement recognition by interaction between active basis model and fuzzy optical flow division. Sci World J 238:234. doi:10.1155/2014/238234

Zachariou V, Klatzky R, Behrmann M (2014) Ventral and dorsal visual stream contributions to the perception of object shape and object location. J Cogn Neurosci 26(1):189–209

Zhong J, Cangelosi A, Wermter S (2014) Toward a self-organizing pre-symbolic neural model representing sensorimotor primitives. Front Behav Neurosci 8:22. doi:10.3389/fnbeh.2014.00022

Chapter 5
Extraction of Grasp-Related Visual Features

A model for visual cue extraction and merging strongly inspired on primate, and especially human, psychophysiology is described in this chapter. This implements the part of the framework of Chap. 4 dedicated to the extraction of object visual features relevant for grasping purposes. The areas and connections of the model of Fig. 4.6 involved in this process are highlighted in Fig. 5.1.

The distance of a target object is estimated, similarly as in the lateral intraparietal sulcus LIP, using proprioceptive vergence data. The advantages of expressing and calculating distances in nearness units are discussed. Object orientation estimation, executed in the posterior intraparietal sulcus CIP, is performed combining binocular (stereoptic) and monocular (perspective) visual data. A theoretical analysis for deriving plausible expressions for slant estimation is accompanied by an implementation with a set of artificial neural networks. The behavior of the system in simulated noisy conditions suggests that the model is faithful to biological reality.

A first interaction between the two streams is implemented at this point. An object recognition module, representing area V4, classifies the target object into one of three basic shapes: boxes, cylinders and spheres. Even though such classification provides no direct information on object size and proportion, it allows to access a basic knowledge about the target shape which helps in the pose estimation process.

The outcome of applying the computational model to a real robotic platform is presented and discussed. The robot is required to observe target shapes of different size and proportion and estimate the features useful for a potential grasping action. The comparison of the obtained results with experiments described in the neuroscience literature confirms the effect of different driving factors on estimation reliability, showing how stereoscopic, perspective and merged estimators behave in different conditions. The same comparison is done for distance estimation obtained from proprioceptive vergence data.

In order to complement the background information provided in the previous chapters, some important concepts regarding cue extraction and integration, object recognition, and artificial vision methods for pose estimation, are given in the next section.

E. Chinellato and A.P. del Pobil, *The Visual Neuroscience of Robotic Grasping*,
Cognitive Systems Monographs 28, DOI 10.1007/978-3-319-20303-4_5

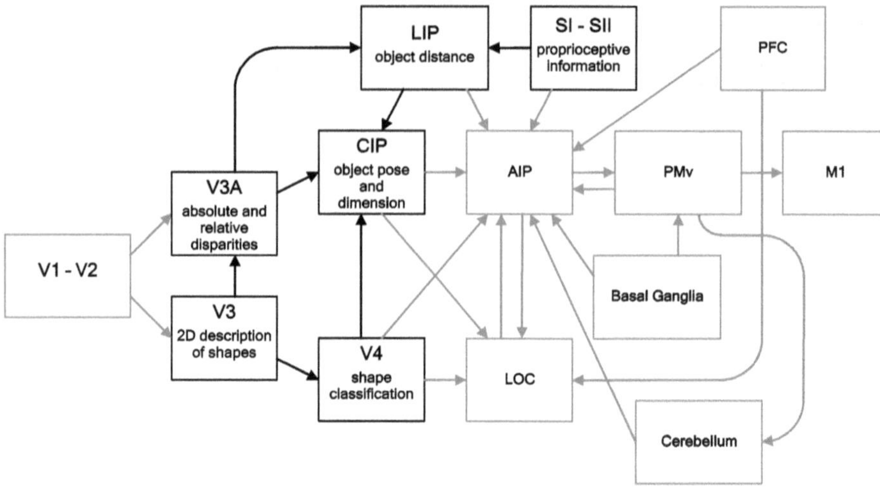

Fig. 5.1 Areas of the model framework involved in extraction of grasp-related visual features

5.1 Extraction and Integration of Object-Related Visual Cues in the Primate Cortex and in Robotics

For both natural and artificial agents, the interaction with the environment requires the ability of estimating distance, size and shape of surrounding objects. Such skill is highly supported by, if not fully dependent on, the use of binocular, or stereoscopic vision (Marotta et al. 1995; Watt and Bradshaw 2003; Bradshaw et al. 2004; Loftus et al. 2004). Binocular vision consists in the contemporaneous acquisition of two different images taken from viewpoints that are always at the same, short distance—the eyes. The process allows to obtain a fast and accurate estimation of object distance, size, motion, through the interpretation of *binocular disparities* (see box).

The difference between the left and right retinal representations of visual features is called *binocular disparity* (Howard and Rogers 2002; Parker 2004). Absolute disparities are simple distances, either horizontal or vertical, between the two retinal positions of the same point-like feature. Various types of higher order disparities can be computed from absolute ones. First order relative disparities represent the difference in disparity between two image points, and thus directly code for feature depth and slant. Horizontal relative disparities are used for estimating object slant about a vertical axis, the most common in nature. Orientation disparities allow instead to calculate slants about horizontal axes (Heeley et al. 2003). Second order disparities are used to estimate object curvature.

Despite its fundamental importance, stereoptic information alone is often not enough, and motion, texture, shading and other cues are used to complement it. Indeed, in each modality the brain seems to efficiently use a large set of different cues at the same time (Norman et al. 1995). Evaluation and integration of all available cues is performed in order to obtain the most likely final estimates. Cue integration is a major principle in the primate sensory cortex, and especially in vision. Visual information is processed in a highly parallel way, different cues for the same stimulus are processed, compared and merged in order to provide increased estimation reliability through redundancy (Landy et al. 1995; Tsutsui et al. 2001). In this section, vision science concepts related to cue generation and integration which help in complementing the review of Chap. 2 are provided.

5.1.1 Feature Extraction

The basic mechanisms of stereoscopic vision have been studied for long time, and are discussed in fundamental works such as Julesz (1971), Marr (1982). Neuronal responses to disparity stimuli in cortical visual areas have also been throughly investigated (Poggio et al. 1988; Cumming and DeAngelis 2001). Disparity detection is a fundamental aspect of visual processing that begins already in V1 and V2 (Gonzalez and Perez 1998; von der Heydt et al. 2000; Thomas et al. 2002; Trotter et al. 2004; Read 2005). It is though from V3 that disparity coding spans areas of the visual field wide enough to provide a proper interpretation of stereoptic information, both in monkeys (Adams and Zeki 2001; Tsao et al. 2003) and in humans (Backus et al. 2001; Welchman et al. 2005; Anzai et al. 2011). For what concerns the processing of higher order disparities, there is a general consensus regarding a prominent role of V3A in representing relative disparities (Backus et al. 2001; Tsao et al. 2003; Rutschmann and Greenlee 2004; Brouwer et al. 2005). An initial, basic perspective processing could also be performed in area V3A (Welchman et al. 2005; Georgieva et al. 2009).

As explained in Sect. 2.3.1, the caudal intraparietal sulcus CIP is dedicated to the extraction and description of visual features suitable for grasping purposes. Its neurons are strongly selective for the orientation of visual stimuli, represented in a viewer-centered way. Selectivity toward disparity-based orientation cues is predominant in macaque's CIP, which neurons are selective for first and second order disparities (Sakata et al. 1998; Endo et al. 2000; Tairaet al. 2000). fMRI studies showed that the human posterior intraparietal sulcus is responsive to disparity-coded orientation, too (Tsao et al. 2003; Rutschmann and Greenlee 2004; Naganuma et al. 2005). On the other hand, many CIP neurons also respond (some exclusively) to perspective-based orientation cues, both in monkeys (Tsutsui et al. 2001, 2005) and humans (Taira et al. 2001).

The evidence suggests that CIP integrates stereoptic and perspective cues for obtaining better estimates of orientation (Tsutsui et al. 2005; Welchman et al. 2005). This sort of processing performed by CIP neurons is the logical continuation of the simpler orientation responsiveness found in V3 and V3A, and makes of CIP the ideal intermediate stage toward the grasping-based object representations of AIP (Sakata et al. 1999; Shikata et al. 2001; Bray et al. 2013; Konen et al. 2013).

Another area that projects to CIP is the lateral intraparietal sulcus LIP, which performs distance and location estimation of target objects. More exactly, according to psychophysiological research in humans (Tresilian and Mon-Williams 2000), what is actually estimated and used in the parietal cortex is the reciprocal of distance, that is, nearness. In the intraparietal sulcus, distance and disparity are processed together, the former acting as a gain modulation variable on the latter (Salinas and Thier 2000; Genovesio and Ferraina 2004). This mechanism allows to properly interpret stereoscopic visual information (Dobbins et al. 1998; Mon-Williams et al. 2000), as described in Sect. 5.2.2.

5.1.2 Cue Integration

Cue integration, or combination, is one of the main working principles of the human sensory systems. Restricting to unimodal cue integration, vision is probably the best example of the complexity reached in the process of getting the best estimate of a stimulus from concurring and often discordant cues. Several models have been proposed for explaining how such best estimate is obtained, but most phenomena can be modeled by a simple linear weighting of concurrent cues, aimed at maximizing the likelihood of the final estimate (Landy et al. 1995). The main underlying principles that allow to achieve this goal seem to be two: cue reliability and cue correlation, or discrepancy (Tresilian and Mon-Williams 2000; Jacobs 2002).

Cue reliability is probabilistic, it depends on environmental conditions, on the estimate itself and sometimes on other, ancillary measures (Landy et al. 1995). Considering the case of interest for our research, i.e. orientation estimation, stereoscopic cues are considered less reliable outside a certain range of disparity, but also at longer distances, being distance in this case an ancillary cue. Often, ancillary cues directly affect the estimation process through gain modulation, such as in the mentioned distance/disparity example (Trotter et al. 1996). This seemingly simple and safe mechanism may nevertheless suffer because of a second-order uncertainty, the problem of assessing the reliability of the ancillary cue itself. In any case, reliability rules have to be learnt by a subject in her/his interaction with the environment, and can be misleading in the case of unusual situations, such as in optical illusions.

The second principle, cue correlation, considers the degree to which concurrent cues conflict or coincide, and gains importance with increasing number of cues. In fact, there is no way to choose between two conflicting cues only on the base of cue correlation, but if a cue is the only one in disagreement with a number of coincident cues, it is very reasonable to consider it untrustworthy. Fortunately, vision systems often provide many cues quite different from each other, so that correlation can be a perfect criterion for weighting the cues in the final estimate (Backus and Banks 1999).

The available models for extraction and integration of visual cues usually focus on very specific aspects, such as disparity responsiveness with changing distance (Lehky et al. 1990; Lehky and Sejnowski 1990), conflicting stimuli (van Ee et al. 1999), maximum-likelihood cue integration (Hillis et al. 2004), temporal integration according to cue reliability (Greenwald et al. 2005), extraction of local surface slant (Jones and Malik 1992). Apparently, no published models on the subject provide details for practical implementation on robotic vision setups.

5.1.3 Object Recognition in the Ventral Stream

As pointed out in Sect. 2.4.1, object recognition in the ventral stream is performed gradually and hierarchically (Grill-Spector et al. 1998; Bar et al. 2001). Recent findings indicate that object recognition is composed of at least two subsequent stages, categorization and identification (Grill-Spector and Kanwisher 2005). In the first stage, an object is classified as belonging to a given class or family of objects, and such process is strikingly fast. The classification delay is so short that there is probably time to feed category information to the dorsal stream, for improving the online estimation of action-related features. This mechanism is represented by the link projecting from area V4 to CIP in Fig. 5.1. As pointed out in Sect. 2.3.1.3, anatomical and functional evidence supports this early integration between the streams (Perry et al., 2014). The second stage of object recognition is proper identification, performed by LOC, in which object identity is recognized within its category.

A second aspect, relevant for modeling purposes, is the method employed by the ventral stream for performing object recognition (Ullman 1996). At least for the first classification stage, visual input is very likely compared to memorized 2D representations (Bülthoff et al. 1995; Orban et al. 2006), and complex objects are identified by composing simpler features (Thoma and Henson 2011). A classification based on 3D representations would require mental rotation, and this can hardly be performed with the quickness observed in the experiments of Grill-Spector and Kanwisher (2005). Moreover, the consistent preference of some "canonical" views during free and classification-oriented object exploration indirectly supports the existence (if not the dominance) of 2D object representations (Blanz et al. 1999; James et al. 2001).

Various biologically inspired methods for object recognition have been developed in computer vision, and different models of ventral stream processing are available (O'-Reilly and Munakata 2000; Riesenhuber and Poggio 2000; Rolls and Webb 2014). Some of them are strongly inspired by neuroscience findings, and use plausible approaches such as radial basis function networks (Pouget and Sejnowski 1997; Deneve and Pouget 2003) or a temporal coherence principle in unsupervised learning (Einhäuser et al. 2005). For the purposes of this work, object recognition is functional to grasping actions, and the interest is not in detailed modeling of ventral stream mechanisms. A simple viewpoint invariant classification is implemented, based on basic 2D global object representations (see Sects. 5.2.3 and 5.4.2).

5.1.4 Orientation Estimation in Artificial Vision and Robotics

Object orientation (or slant) estimation is a common, and difficult, problem in artificial vision (Lippiello et al. 2006). Nevertheless, no research works similar to the proposed approach are available in the literature.

A detailed overview of existing techniques for pose estimation can be found in Goddard (1998). The available approaches differ depending on the type and location of the sensors, the illumination requirements, the object or scene feature on which the pose is calculated, the relative motion between robot and object. Sometimes, noise sources and uncertainty factors are modeled in an attempt to improve the robustness and accuracy of the results. Among various methods, geometry or model based techniques are most common. These methods use an explicit model for the geometry of the object in addition to its image in determining the pose. The object is modeled in terms of points, lines, curves, planar surfaces, or quadric surfaces (Rosenhahn et al. 2004). Methods of this kind have been proved useful even with moving targets (Lippiello et al. 2001). Often, the use of markers substitute explicit modeling (Gehrig et al. 2006). These techniques can be combined with others, where appearance based methods are used for the rough initial estimate and followed by a refinement step using model based technique (Ekvall et al. 2003). In Peters (2004) the rough initial estimate is determined on the viewing hemisphere as an initial guess, and then also refined. A model based approach can also be connected with range images, for example matching a 3D model to a range representation of the scene (Germann et al. 2007). The managing of range data is anyway quite different from vision research, and works which locate parallel surfaces to grip from range images, such as Weigl et al. (1995), are interesting but unrelated to the current approach.

For what concerns stereo slant estimation inspired on human physiology, Ferrier (1999) describes a method based on disparities which makes use of a model for computing orientation of features. With the support of camera calibration, which is not used in this work, they obtained similar results. Regarding the integration of stereoptic and perspective cues in artificial vision, although the idea is not novel (Clark and Yuille 1990), there are very few robotic platforms that make use of both visual cues at the same time. For example, in Saxena et al. (2007) a vision system is trained to estimate scene depth through monocular data using supervised learning, and a joint monocular/binocular estimator is generated. The authors show that integration of monocular and stereopsis data performs better than either cue alone. Other works, focused on object tracking (Taylor and Kleeman 2003) and on visual servoing (Kragic and Christensen 2001), perform cue integration, but their visual analysis is model-based, and their goal is feature matching and not feature extraction.

5.2 A Model of Distance and Orientation Estimation of Graspable Objects

This section introduces a proposal for distance estimation based on proprioceptive vergence data, and two different orientation estimators obtained from stereoscopic and perspective cues. A hierarchical approach to object classification is also presented.

5.2.1 Distance Estimation Through Proprioceptive Data

The distance of a fixated object from the viewer can be estimated by either retinal and/or proprioceptive cues, *accommodation and vergence* (see box).

> The movement performed by the eyes in order to converge on a given visual target is called *vergence*, or *convergence*. The resultant angle is called *vergence angle*. The adaptation of the shape of the eye crystalline lens in order to change the eye focus is called *accommodation*. Accommodation and vergence are both directly related to the distance of the visual target, and are linked by a reflex. In distance estimation, accommodation and vergence are preferably used when retinal data are not available or considered not reliable, and for short distances (Mon-Williams and Tresilian 1999; Tresilian et al. 1999).

The relation between distance and vergence angle γ_P is simple and depends only on the interocular distance I, which is constant (see Fig. 5.2). The distance d between the fixated point P and the cyclopean eye O, middle point between the two eyes, is given by:

$$d = \frac{I}{2\tan(\gamma_P/2)} \tag{5.1}$$

Psychophysiological experiments (Tresilian and Mon-Williams 2000) suggest that distance estimation is most probably performed in the human brain using *nearness* units instead of distance units. Nearness is the reciprocal of distance, and a point at infinite distance has 0 nearness. The nearest distance at which vergence can be maintained, called the *near point of vergence*, is usually between 60 and 70 mm (Brautaset and Jennings 2005). Average interocular distance for adults is considered to be between 63 and 65 mm, and thus approximately coincident with the near point of vergence. Setting $I = d$ yields a maximum vergence angle of: $\gamma = 2 \cdot \arctan(I/(d \cdot 2)) = 2 \cdot \arctan(1/2)) = 0.927\text{rad} = 53°8'$. The following expression for computing nearness from vergence hence produces nearness values between 0 (for $\gamma_P \to \infty$) and 1 (for $\gamma_P = 0.927$ rad, the maximum vergence angle):

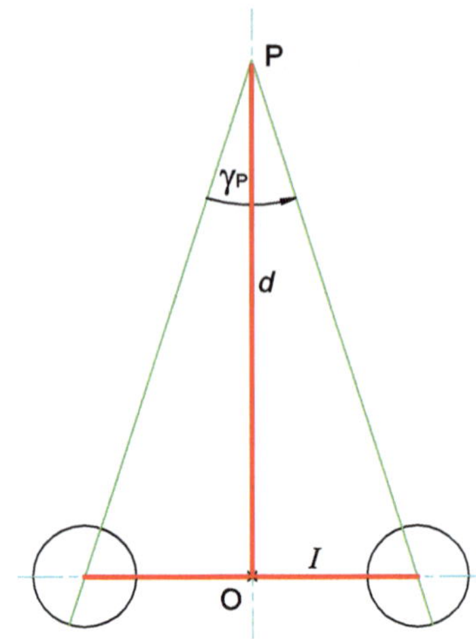

Fig. 5.2 Relation between vergence angle γ_P and distance d. I is the interocular distance, O the position of the cyclopean eye

$$nearness = 2\tan(\gamma_P/2) \tag{5.2}$$

Such measure is more precise for close distances, and thus especially suitable for dealing with objects in the peripersonal space. Moreover, it is based on the relation between I and the near point of vergence, and does not depend on any constant or auxiliary measures.

Two radial basis function (RBF) networks were designed, for learning the association between vergence and nearness and between vergence and distance. The results can be seen in Fig. 5.3. On the top left, the distance/vergence curve corresponding to (5.1) is shown. Equation (5.2) between vergence and nearness is depicted on the top right of the image, and the corresponding learnt curve appears on the bottom right of Fig. 5.3 (lighter, dashed curve). The reciprocal of the learnt relation is finally depicted on the bottom left, where it can be compared with the true mathematical relation (they practically coincide). In this simplified example, to obtain a similar performance in the estimation of distance, the distance/vergence network requires 11 RBF units, while the nearness/vergence net requires only 4 neurons. This is not surprising, considering the approximate linearity of the relation vergence/nearness, and considering that the brain often employs an economy principle, minimizing the resources required to perform a given task. In the current model, object distance is represented in nearness units, and is used in the next section to modulate the effect of disparity on orientation estimation.

5.2.2 *Object Orientation Estimation Through Retinal Data*

For estimating object pose, humans combine estimators provided by different visual and proprioceptive cues, both binocular (mainly horizontal and gradient disparity

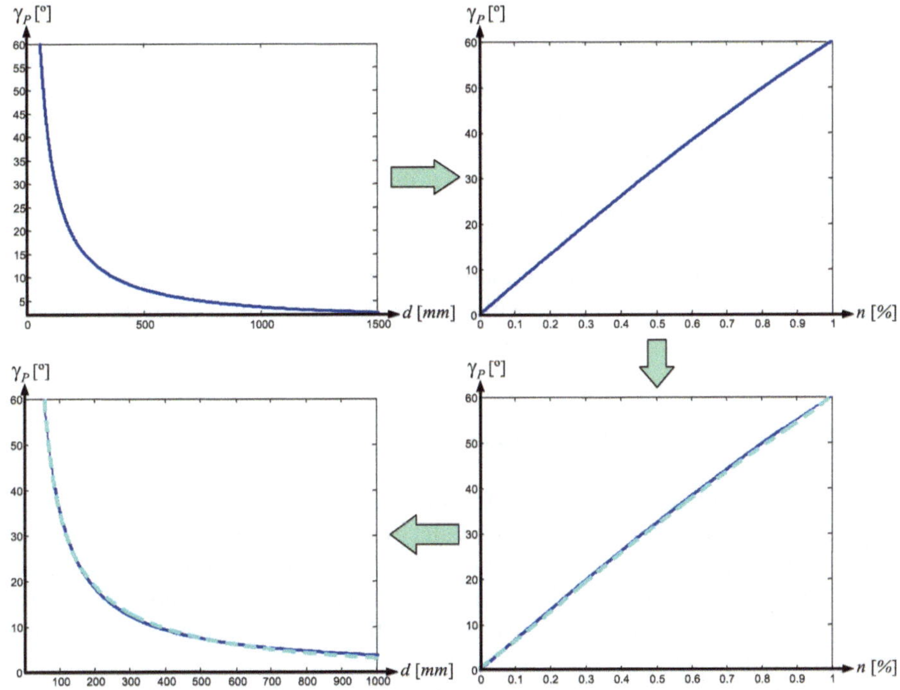

Fig. 5.3 Distance to vergence and nearness to vergence relations: theoretical equations (*upper diagrams*) and learnt *curves* (*superposed dashed lines in the lower diagrams*)

cues) and monocular (texture or edge perspective cues) (Backus et al. 1999). In this section, the problem of orientation estimation according to different cues is analyzed. First, a couple of expressions both neurologically plausible and useful for a practical implementation are derived. Next, the neural network architecture implemented for the solution of the problem is introduced. Finally, some results which allow to discuss the theoretical and practical implications of the proposed approach are described.

The proposed object orientation estimation process makes use of simple visual information for achieving a geometric 3D selectivity similar to that observed in neuroscience studies. The goal is to develop a modular computational structure, composed of various estimators, which makes use of proprioceptive and retinal cues in order to obtain the geometrical parameters needed for grasp planning. This approach differs from related research (e.g. Jones and Malik 1992) in that it builds upon retinal data: instead of using pixelated images and projective matrices, the only inputs are retinal angles and proprioceptive eye data. The center of the coordinate system is the cyclopean eye, as for humans.

For orientation and basic shape discerning, the approach relies upon one monocular information source, that is, perspective under the assumption of edge parallelism, and one kind of binocular information, width disparity. As explained in the previous section, these data are coded by visual areas V3 and V3A and combined in the posterior intraparietal cortex CIP. One basic assumption is that objects recognized as boxes or cylinders have actually straight, parallel edges, and are laying on a horizontal table. This is very plausible from a neuropsychological point of view, as the primate brain is actually "programmed" to better assess vertical and horizontal edges, most common in nature. Indeed, experiments on monkeys (Tsutsui et al. 2001) and humans (Brouwer et al. 2005) have shown that, even for purely perspective pose estimations, a frontoparallel trapezoid is usually interpreted as a rectangular shape slanted in depth.

Next, we analyze the sort of computation performed by the human brain during orientation estimation, in the binocular and in the monocular case, and propose plausible transfer functions to obtain estimators from simple retinal angles.

5.2.2.1 Stereoscopic Slant Estimation

In Fig. 5.4a a viewing scene is seen from above: object PQ of length l is slanted about a vertical axis with an orientation θ. Its extreme P is the fixation point, placed straight ahead from the cyclopean eye (in this way γ_P corresponds to the vergence angle). All α angles represent the retinal projections of points P and Q on the left and right eyes, I is the interocular distance, ψ_Q the binocular separation of points P and Q (being $\psi_P = 0$).

The change in slant of segment PQ as point Q moves on the xz reference system can be observed in Fig. 5.5. In the graph, the position of P is fixed at ($x_P = 0$, $z_P = 30$). The dot represents θ for Q positioned as in Fig. 5.4a.

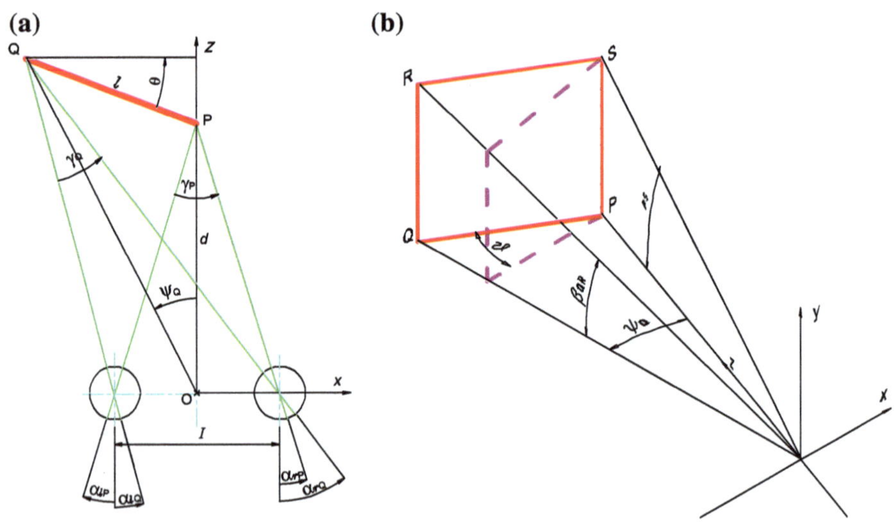

Fig. 5.4 Schemes for deriving slant from stereopsis and perspective. **a** Stereopsis. **b** Perspective

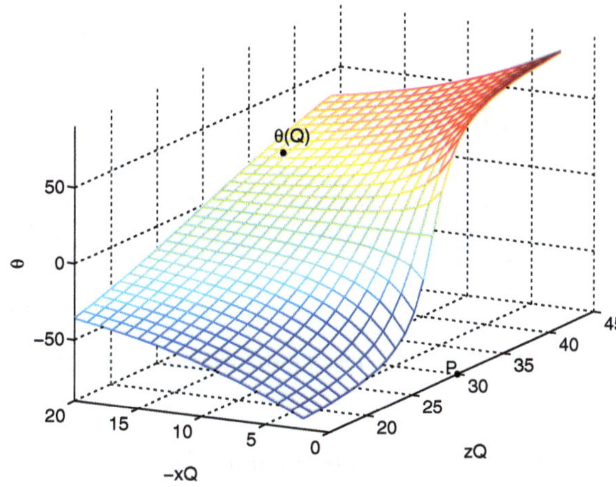

Fig. 5.5 Slant θ as a function of the position of point Q in the xz space

The horizontal slant θ of an object can be computed only from retinal angles using the following expression, which can be derived from Fig. 5.4a:

$$\tan\theta = \frac{(\tan\alpha_{rQ} - \tan\alpha_{lQ}) - (\tan\alpha_{rP} - \tan\alpha_{lP})}{\tan\alpha_{lP}\tan\alpha_{rQ} - \tan\alpha_{lQ}\tan\alpha_{rP}} \tag{5.3}$$

Reminding that P is the fixation point, so that $\alpha_{lP} = -\alpha_{rP} = \gamma_P/2$, the equation can simplified in this way:

$$\tan\theta = \frac{1}{2\tan(\gamma_P/2)} \cdot \frac{(\tan\alpha_{rQ} - \tan\alpha_{lQ}) - (\tan\alpha_{rP} - \tan\alpha_{lP})}{(\tan\alpha_{rQ} + \tan\alpha_{lQ})/2} \tag{5.4}$$

Recalling (5.2) and the definitions in the disparities text box, this relation can be expressed by using only quantities that are actually computed in the visual brain areas:

$$\tan\theta = \frac{1}{nearness} \cdot \frac{relative\ disparity}{separation} \tag{5.5}$$

Separating θ, a biologically plausible stereoptic orientation estimator $\hat{\theta}_S$ is obtained:

$$\hat{\theta}_S = \arctan\frac{relative\ disparity}{nearness \cdot separation} \tag{5.6}$$

The interpretation of (5.6) is that the component due to disparity (the fraction relative disparity/separation, which is also called disparity gradient) is modulated by the viewing distance (or nearness), as indicated by neuroscience research. Nearness is probably computed from proprioceptive data, as discussed in the previous section.

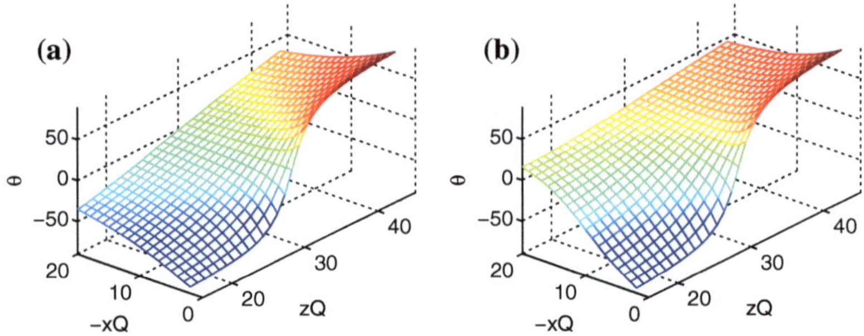

Fig. 5.6 Distorted estimation of θ (**b**) obtained by approximating $\tan\alpha$ with α

The separation which appears in the formula is binocular, referred to the cyclopean eye.

Other works (Banks et al. 2001; Howard and Rogers 2002) propose expressions similar to (5.6), starting from slightly different assumptions. It is important to point out though that common approximations found in the literature carry to unacceptably wrong estimations in most cases. An example of this can be seen in Fig. 5.6, where an exact reproduction of Fig. 5.5 (Fig. 5.6a), obtained with (5.6), is compared to a distorted one (Fig. 5.6b), obtained with the same equation in which the common solution of approximating the exact value of $\tan\alpha$ with α is employed. The comparison demonstrates that for a real application, and a faithful model, exact expressions need to be used.

5.2.2.2 Perspective Slant Estimation

The slant of an object can be estimated using only monocular data, as depicted in Fig. 5.4b, in which the origin of the axes is one of the eyes. The frontal object face is considered as rectangular, exploiting the reasonable assumption of parallelism and equality of opposite edges (PS and QR in the image). The angles β in the figure represent the vertical retinal angles associated to such edges. The function which leads from retinal angles to orientation estimation can be derived from the draw, and can be referred entirely to either the left or the right eye:

$$\tan\theta = \frac{\tan\beta_{QR}}{\tan\beta_{PS}\sin\psi_Q} - \frac{1}{\tan\psi_Q} \tag{5.7}$$

In this case the monocular separation is: $\psi_Q = (\alpha_Q - \alpha_P)/2$.

Approximating $\sin\psi_Q$ to $\tan\psi_Q$, which is plausible for reasonably small separations, a formula for perspective estimation of θ is obtained:

$$\hat{\theta}_P = \arctan\frac{perspective\ disparity}{separation} \tag{5.8}$$

Perspective disparity is the quantity $\frac{\tan \beta_{QR}}{\tan \beta_{PS}} - 1$, which represents the proportion between the projected sizes of edges QR and PS. Therefore, again, the estimator depends on a separation factor and a disparity factor, this time monocular.

Equations (5.8) and (5.6) will be used, both separately and merged, for simulated (Sect. 5.3) and real orientation estimation on a robotic setup (Sect. 5.4).

5.2.3 Hierarchical Object Classification

The approach to object classification proposed in the model is composed of a three stages process. These stages are initial shape classification, proper object recognition and actual identification of a known object.

1. **Shape classification**. In this stage the target object is classified into one of a number of known classes. For example, a bottle would be classified in the class of cylinders. Simple visual information such as shape silhouette or a basic topographic relation between object features is enough for this task. No actual data regarding the size and the proportion of the object are considered. Nothing is inferred at this point about object composition, utility, meaning. The information recovered at this stage is used by early areas of the dorsal stream in order to estimate the size and pose of the object.
2. **Object Recognition**. Actual object recognition is the goal of this stage. The target object is identified as if the task was to name it. What was a general cylindrical shape in the previous stage is now identified as a bottle. Additional conceptual knowledge is thus added to the previous basic information. Composition, roughness, weight of the object can be inferred if not known for sure. The object proper use in different tasks is also recalled at this point. Object recognition directly affects the process of grip selection, providing a bias toward grasp configurations better suited to the object weight distribution, possible friction and common use.
3. **Object Recall**. In this final stage, a subject recalls a single well-known object which was encountered, and possibly grasped, before. Going back to the cylinder example, here it can be recognized as a wine bottle recently bought, and thus previously known and dealt with by the subject. Compared to the previous one, this stage adds security to the estimation of the object characteristics. To recognize an object as a bottle helps in estimating its weight, whilst to identify a previously encountered bottle provides an exact value of that weight.

In all stages, the classification process has to be viewpoint invariant. A very important issue is that object classification and recognition is always a gradual process, not a binary one, and each classification is accompanied by a confidence value, necessary to clarify its reliability. Any classification having a low confidence should be used prudentially, and if no class or object are clearly identified the system should rather provide a failed classification answer, to clarify that the situation is uncertain and needs further exploration. Feedback from execution outcome can later be used to complete and improve the world knowledge in these situations.

5.3 Neural Network Implementation of a Multiple Cue Slant Estimator

A neural architecture for estimating the orientation θ of a target object according to the concepts described in the previous section has been implemented. The framework includes two sets of neural networks, for stereoscopic estimation and for monocular estimation based on perspective data.

5.3.1 Neural Network Estimators

The whole framework of the neural network implementation is depicted in Fig. 5.7, where the nets are associated to the brain areas that probably perform their functions. Apart from nearness estimation, implemented with the RBF network described in Sect. 5.2.1, all networks are feedforward backpropagation, trained with the Levenberg-Marquardt algorithm. Four neural networks constitute the module for orientation estimation based on stereopsis: they compute nearness, relative disparity and separation (two nets represented as a unique one in the scheme), and the final estimate of $\hat{\theta}_S$ from the outputs of the previous three networks (according to expression (5.6)). The module for orientation estimation based on perspective makes use of two networks for computing the two components of (5.8), and a third for the final calculation of $\hat{\theta}_P$.

5.3.2 Merging the Estimators

Following the insights provided by the neuroscience literature, the final orientation estimator is computed by combining the stereoscopic estimator $\hat{\theta}_S$ and the perspective

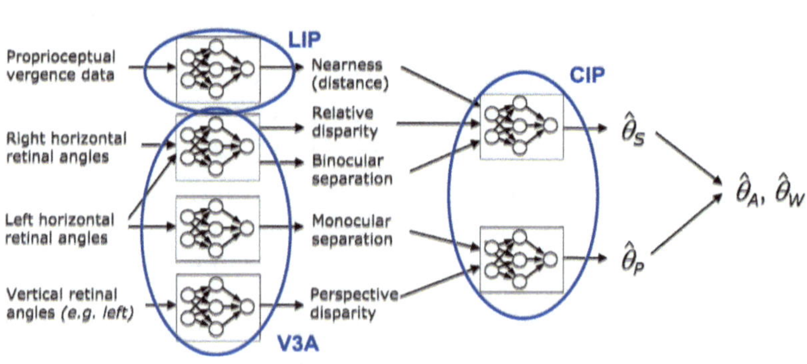

Fig. 5.7 Scheme of the neural network architecture for slant estimation

estimator $\hat{\theta}_P$. A first merging can be done through a simple average of the two output values:

$$\hat{\theta}_A = \frac{\hat{\theta}_S + \hat{\theta}_P}{2} \tag{5.9}$$

Looking for better performances and improved biological plausibility, the two principal driving factors for cue merging, correlation and reliability, have been taken into account. In the present case, cue correlation could not be used, as only two different estimators are available. The chosen solution was thus to experimentally simulate cue combination using only cue reliability. The stereoptic and perspective estimators were trained to learn how their reliability changes in different conditions. According to the literature, the driving factors for the accurateness of orientation estimation are distance and orientation itself (this is not a contradiction: the estimated value can be used as output and, at the same time, as reliability index for the estimation). In fact, although it is known that stereopsis quickly looses its reliability with distance, here the interest is on the near space defined by the arm reaching distance, within which the variation of distance affects the two methods in similar ways. For this reason, the focus is rather put on the effect of orientation, and the goal is to devise a merging method that optimizes the weights given to the two estimators when changing the estimated value of θ.

How the human brain can predict cue reliability is still a matter of debate. Nevertheless, it has been shown that stereoscopic and perspective cues are actually weighted through a maximum-likelihood process (Knill 2007). To emulate this process, the error patterns obtained with the estimation methods alone were saved, and used to generate a joint estimator which is a weighted average of the original ones:

$$\hat{\theta}_W = w_S\hat{\theta}_S + w_P\hat{\theta}_P \tag{5.10}$$

In (5.10), w_S and w_P are functions of θ computed in the following way:

$$w_S = \frac{SSE_P}{SSE_S + SSE_P}; \qquad w_P = \frac{SSE_S}{SSE_S + SSE_P} \tag{5.11}$$

where SSE_S and SSE_P are the previously learnt summed squared errors of stereopsis and perspective respectively.

5.3.3 Results of the ANN Simulation

In principle, the neural network implementation allows to achieve any arbitrary precision in the estimation. The study of estimators reliability can thus be done either in the real world or simulating the effect of natural imprecisions introducing stochastic variability in the computation. Before the implementation on a real robotic setup, a

Table 5.1 ANN slant estimation results for different estimators

Method	Estimator	Error (°)
Perspective	$\hat{\theta}_P$	4.49
Stereopsis	$\hat{\theta}_S$	4.17
Simple average	$\hat{\theta}_A$	3.05
Weighted average	$\hat{\theta}_W$	2.93

simulation was performed to study the effect of noise on slant estimation performed with stereoptic and perspective methods. With this purpose, random noise was added to all retinal angles which constitute input values for the nets. The error representing the average difference between estimation and true value was calculated all over the test space ($10° < \theta < 70°$, $450\,\text{mm} < d < 850\,\text{mm}$).

In Table 5.1 the improvement of joining the two estimators in this way can be observed. For comparison, consider that the nets were trained so that the average error of the two original estimators $\hat{\theta}_S$ and $\hat{\theta}_P$ before the introduction of noise was less than one degree. Although stereopsis seems to suffer less from the insertion of noise, the contribution of both perspective and stereoptic predictors is very important for improving the final result. In fact, the weighted average $\hat{\theta}_W$ allows to obtain an improvement of almost 30 % on the best single cue estimator $\hat{\theta}_S$, suggesting that the combination of different cues is the best solution for pursuing a reliable estimation. Even the simple average $\hat{\theta}_A$, that can be used a priori without exploiting previous experience, improves the $\hat{\theta}_S$ performance by more than 25 %. The performance difference between $\hat{\theta}_W$ and $\hat{\theta}_A$, which is rather small in this example, sensibly increases especially in the most extreme situations, when one of the estimators is much better than the other, and the simple average would not take this aspect into account.

Experiments with human subjects tell that distance, as an ancillary cue, and slant itself are the two most important driving factors for slant estimation reliability. Figure 5.8, taken from Hillis et al. (2004), depicts the precision of two orientation estimators, perspective and disparity based, as a function of distance and slant. With increasing distance, both estimators become less reliable, but the stereoscopic cue (blue) is clearly more affected. The effect of orientation is more complex. Perspective methods are more sensitive and precise for pronounced slants, that generate higher differences in vertical disparities. At long distances, disparity methods also prefer high slants. On the contrary, for the short distances typical of grasping actions their error is minimum for low slant values, which grant higher binocular disparities.

To check if this pattern of behavior could be reproduced in the simulation, the estimators accurateness was plotted as a function of distance d and as a function of orientation θ. The outcome can be observed in Fig. 5.9, in which the error in stereoptic and perspective estimation is plotted against orientation (Fig. 5.9a) and distance (Fig. 5.9b). The similarity of the obtained results to what is described in the literature is remarkable, as can be observed by comparing the corresponding ranges of Figs. 5.8 and 5.9. Notice that, being Fig. 5.8 symmetrical with respect to slant, in Fig. 5.9a only positive slants are plotted, and Fig. 5.9b considers just reachable

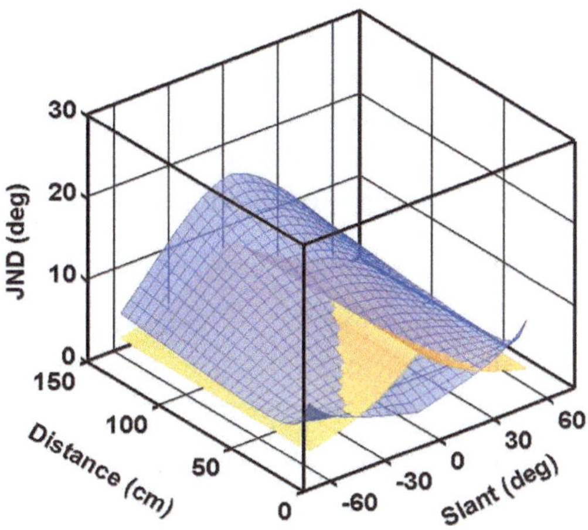

Fig. 5.8 Precision of texture (*orange*) and disparity (*blue*) cues as a function of distance and slant. JND is the Just Noticeable Difference, corresponding to the smaller detectable slant variation. From Hillis et al. (2004)

distances, up to 850 mm. The proposed model looks thus appropriate to reproduce the behavior of stereoptic and perspective estimators.

The second effect that could be reproduced is the improved performance obtained through a maximum likelihood merged estimator in which cues are weighted according to their reliability (experimentally learnt), as explained in Sect. 5.1.2. The better results obtained with the weighted estimator θ_W can also be observed in Fig. 5.9.

The implemented neural architecture hence constitutes an orientation estimator both biologically plausible and practically reliable. The quantities used are employed by the human visual system, but also computationally useful for artificial implementation (e.g. retinal angles). The proposed equations for computing orientation from stereopsis and perspective are plausible transfer functions useful to model the estimation process. The trained neural networks are someway emulating the behavior of modules pertaining to higher visual brain areas. Indeed, inputs and intermediate results represent quantities that have been observed and measured, and are part of real brain processes (Welchman et al. 2005). This suggests that functions (5.6) and (5.8) are plausible models for stereoptic and perspective slant estimation in the human cortex.

The simulation results indicate that the proposed approach can be suitable for improving the reliability of a real application. The next immediate step is the practical experimentation on a robotic platform.

5.4 Robotic Validation

Orientation and pose estimation are very complex problems in machine vision, especially when the goal is to develop a reliable robotic system which makes use of visual estimates to interact with the environment, such as in object grasping actions (Wandell 1995; Trucco and Verri 1998). In this section, the computational method

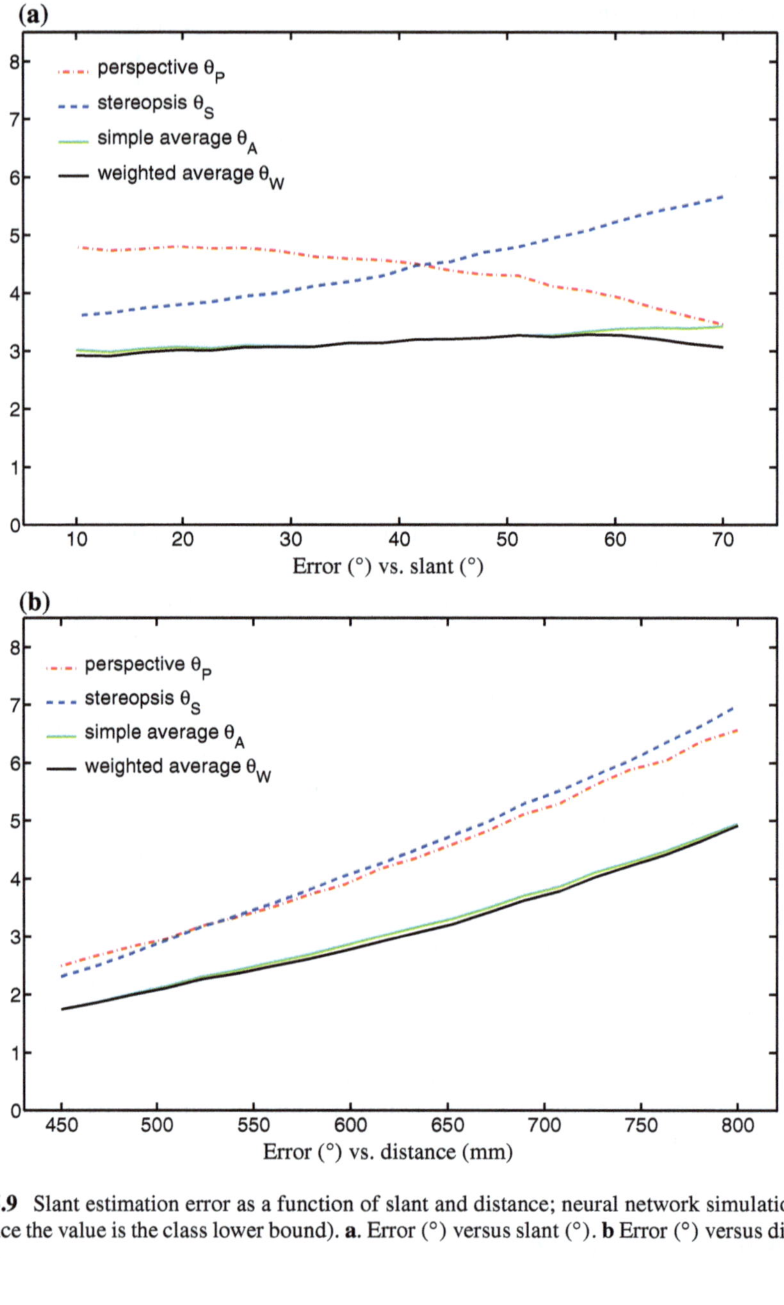

Fig. 5.9 Slant estimation error as a function of slant and distance; neural network simulation (for distance the value is the class lower bound). **a.** Error (°) versus slant (°). **b** Error (°) versus distance (mm)

described above is implemented on a robotic setup. The goal is to obtain an orientation estimator robust and reliable enough to be used in vision-based robotic grasping. A number of different experiments to verify how the ideal results change when the model has to face the uncertainties of the real world are executed. As a second goal, the implementation tries to reproduce the effects obtained with the ANN simulation with real experimental data, and hence further validate the model.

5.4.1 Robotic Setup

The robotic setup, shown in Fig. 5.10a, consists of a seven degrees-of-freedom (DOF) Mitsubishi PA-10 arm endowed with a Barrett Hand and a JR3 force/torque and acceleration sensor mounted at the wrist, between hand and arm. A stereoscopic, black and white camera Videre Design is coupled to the wrist, eye-in-hand style (Fig. 5.10b). This configuration allows for controlled movements of the vision system without the need of a pan-tilt-vergence robotic head.

The Barrett Hand (see schema in Fig. 5.11) has three-fingers with a total of four controllable degrees of freedom. Each finger possesses two joints which are driven by a single motor. The controlled variables are thus the three finger extensions e_1, e_2 and e_3. The fourth degree of freedom controls the opening angle θ of fingers 2 and 3, which are symmetrically placed on either side of finger 1, the *thumb*, which is fixed. When fully abducted, for $\theta = 0°$, fingers 2 and 3 oppose the thumb, when adducted ($\theta = 180°$) they flex in parallel to the thumb.

As it can be observed in Fig. 5.10b, the hand fingertips are equipped with arrays of pressure sensors, designed and implemented by Weiss Robotics (Weiss and Wörn 2004). The sensors are 8×5 cell matrices that cover the inner parts of the distal phalanxes of the fingers. Each sensor is able to detect a complete two dimensional force profile by the use of a homogeneous material connected to an adequate electrode matrix.

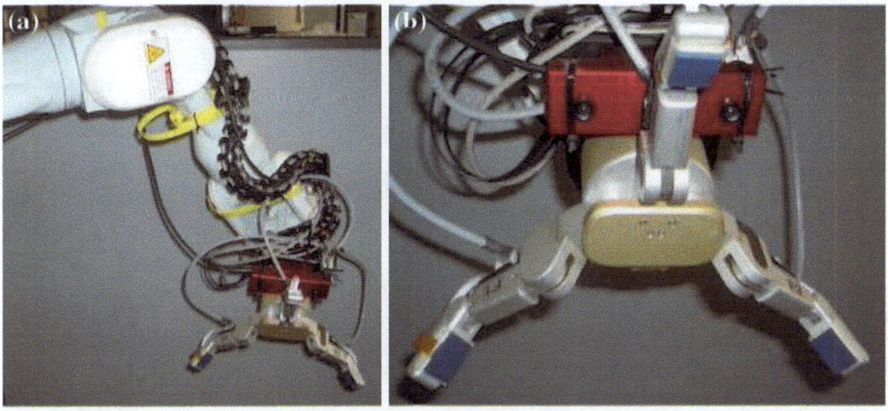

Fig. 5.10 Robotic setup with arm, hand and stereoscopic camera. **a** Robotic arm and hand. **b** Detail of hand with stereo camera

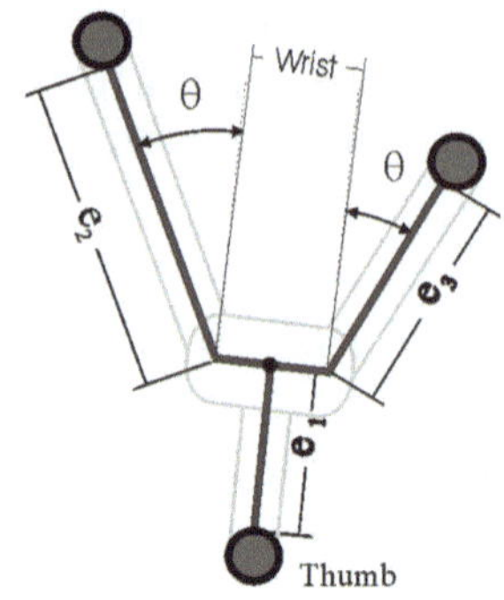

Fig. 5.11 Barrett hand kinematics

The robot world is a dark environment in which clear shapes are placed on a table at variable positions and orientations (see Fig. 5.12). The range of possible positions are those that allow to view the object and also keep it at reaching distance for the hand. Using the estimators previously introduced, the system is able to estimate distance, pose and size of objects without using explicit models, but only common knowledge regarding basic shapes it recognizes, such as the assumption of edge parallelism.

The grasping action begins with the stereo camera facing straight ahead, and having an object in its field of view. Both left and right images are continuously binarized and the object contour tracked. The choice of object and background color is driven by the need of keeping image processing as fast and lean as possible. The point in the image having minimum y coordinate, called P, is selected as reference and starting point for the contour, and one of the images is centered on it. Let us

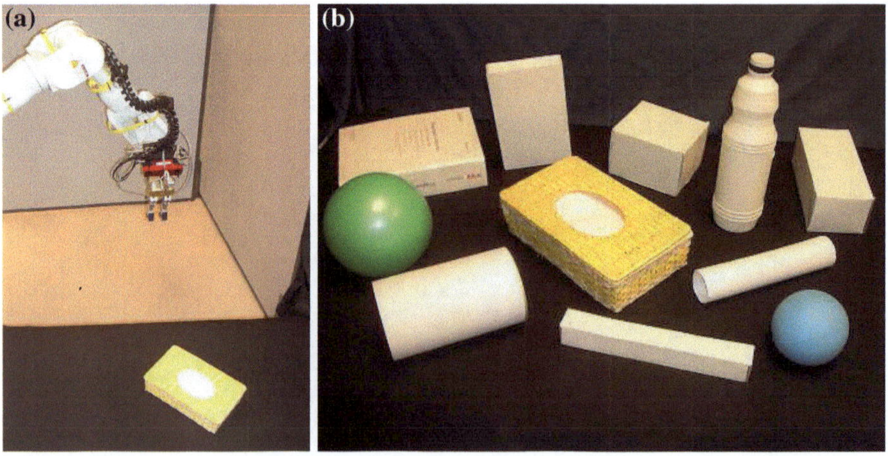

Fig. 5.12 Workspace with robot fixating an object and possible target objects as seen from the robot camera. **a** Robot at fixation position. **b** Workspace with examples of target objects

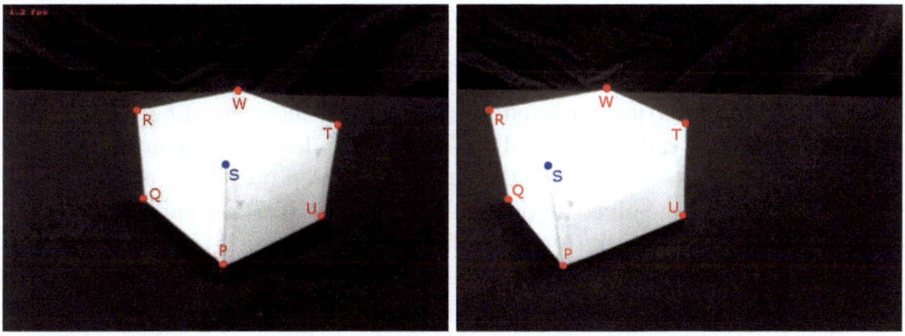

Fig. 5.13 *Left* and *right* object images from the initial position, with labels of detected corners

assume from now on that processing begins from the left eye, hence point P is centered on the left image (see Fig. 5.13 in Sect. 5.4.3).

The images at this position are processed by two parallel modules, one concerned with classifying the object in a number of known categories, the other dedicated to pose estimation. The first module, emulating the processing of the medium ventral stream, makes use of a global visual representation of the object in order to perform a viewpoint invariant classification. The second module integrates different cues for estimating object distance, size and pose.

5.4.2 Object Classification Experiments

The object classification module has to categorize objects seen from different poses and distances. With this purpose, it has to consider object images globally, rather than focusing on local features. The goal is to classify an object as pertaining to one of three known object classes: parallelepipeds (boxes), cylinders and spheres. This has to be done using only a couple of stereo images, without changing the viewpoint. Moreover, it is important to retrieve a value measuring the confidence in the classification, represented by the percentage of likeliness assigned to each class. Two different approaches were tested, using the extracted object contour as a silhouette of the object.

The first tested method consisted in computing a chain code of the contour, which constitutes a representation that is invariant with respect to size and distance, while maintaining the feature topology necessary to identify the object. The chain was generated extracting a pre-defined, finite number of points regularly spaced along the contour, starting from P. The code c_i corresponding to point P_i is thus the following, normalized so that the range is $[-1, 1]$:

$$c_i = atan\left(\frac{y_i - y_{i-1}}{x_i - x_{i-1}}\right) \tag{5.12}$$

The number of points to use can be chosen according to the application. The selection of 20–30 points gave the best results. The chain codes of different objects from different points of view constituted the training data. A probabilistic neural network was used to classify the object in one of the three classes. For training, 10 different objects were used, each seen from 19 different positions (apart for the spheres).

This solution did not provide the required behavior. In fact, results on training objects (form different viewpoints) and on different test objects gave recognition success very close to 100 %, but test objects were often misclassified. Moreover, even in the wrong cases, confidence was always very high, often above 98–99 %. The conclusion is that the method is very good in recognizing known objects, but not in generalizing. The sequential order of different object features, like straight and curved segments, or corners, would be enough for classification. Instead, the chain code representation takes into account and hence classify objects also according to the feature length, distinguishing for example a short cylinder from a long one. Moreover, classification should be much more shaded, with confidence percentages not always close to 100 %. As justified in Sect. 5.2.3, a missed classification due to high uncertainty is preferred to a wrong categorization.

For these reasons, a different classification method was tested, based this time on the curvedness of objects. This method is based on only one index representing each object, the curved fraction of its contour, ratio between the length of its curved features and the total contour length. For the shapes in use, experimental data showed that parallelepipeds, cylinders and spheres normally possess linearly separable curvedness values. The classification process begins with a training phase during which the system is presented with five different boxes (B), three cylinders (C) and two spheres (S), again from 19 viewpoints distributed along a 90° range. Average curvedness values μ_K and corresponding standard deviations σ_K are calculated for the three classes, $K \in \{B, C, S\}$.

Given a test point c_i, the curvedness coefficient of object i, its degree of membership m_{iK} to class K is computed as the reciprocal of the relative distance to the class center:

$$m_{iK} = \frac{\sigma_K}{|c_i - \mu_K|} \tag{5.13}$$

At this point, classification percentages for the three classes $K = B, C, S$ are given by:

$$p_{iK} = \frac{m_{iK}}{m_{iB} + m_{iC} + m_{iS}} \tag{5.14}$$

As explained above, a missing recognition response is better than a misclassification. To favor the former over the latter, a high confidence value of 70 % is required to assign the object to any class. If no class reach this value, the object is not classified. In such cases, only distance and approximated center of mass (that is in reality the centroid of the visible 2D silhouette) can be estimated and used for grasping.

An exception is the case of uncertainty between boxes and cylinders. If the sum $p_{iB} + p_{iC} > 70\,\%$, then the object is classified in the less restrictive class, i.e., as a cylinder. For cylinders, only one face can be computed for slant estimation, while for boxes two visible faces are used. A misclassification of a cylinder as a box would thus provide a wrong orientation estimation, whilst a misclassification of a box as a cylinder would just imply that some available information is not used.

Classification results for objects in the training set are provided in Table 5.2. Cases of misclassification are highlighted in bold whilst uncertain cases are marked in italics. For the training set, only two problematic cases are identified, both for cylinders seen from a 0° angle (objects 5 and 6). It is not surprising that this is a difficult condition for the recognition system, as the contour provides limited if any information on curvature, and more elaborate methods which take into account shading would be required for proper classification.

Classification results for test objects are given in Table 5.3. Most cases of missing classification regard the problem observed for the training set. Cylinders seem to be difficult to recognize, especially for extreme viewing angles, in which their silhouette appears as a rectangle or as a circle. Nevertheless, the prudential decision of assigning the object to class C in case of uncertainty between box and cylinder, works in nearly all conditions, and provide reliability to the whole pose estimation system. Only objects 14 and 16 from the 0° viewpoint are finally misclassified, the first as a sphere and the second as a cylinder. Object 18 cannot be clearly put in any of the three classes, but it has one face that can be used for slant estimation, as cylinders, hence its classification as a cylinder is the most appropriate from a practical point of view.

5.4.3 Object Pose and Distance Estimation

5.4.3.1 Orientation Estimation

For what concerns pose estimation, this requires the extraction of features as those used in the model. For this reason, a number of salient points on the contour have to be extracted. Object classification biases this process, modeling the influence of ventral stream data on dorsal stream processing, and the contextual nature of 3D perception (Todd 2004).

For boxes, the salient points usually correspond to the object corners (Fig. 5.13). Object faces are not segmented separately, so the number of detected corners ranges from 4 to 6 depending on point of view and object pose. Possible missing points are added according to the shape class. For example, if a box is detected by the classification process and, due to a bad perspective position only five points of the contour are chosen as corners, the sixth will be set according to simple geometric considerations. For a cylinder, only four points are necessary, as those on the curved parts of the contour are not used. For spheres only centroid and apparent diameter are computed.

Table 5.2 Object classification percentages for different slants; training shapes

#	Object	Class	0°	30°	60°	90°
1		Box	98.16	86.63	84.81	94.90
		Cylinder	1.64	11.57	13.12	4.44
		Sphere	0.20	1.80	2.07	0.66
2		Box	92.97	85.90	84.81	91.22
		Cylinder	6.13	12.19	13.12	7.63
		Sphere	0.90	1.91	2.07	1.15
3		Box	93.94	84.81	84.84	87.25
		Cylinder	5.29	13.12	13.10	11.04
		Sphere	0.77	2.07	2.06	1.71
4		Box	99.87	86.88	84.81	99.23
		Cylinder	0.11	11.36	13.12	0.68
		Sphere	0.02	1.76	2.07	0.09
5	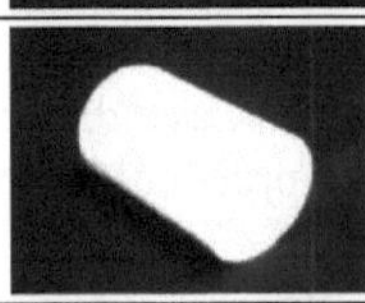	Box	**86.20**	0.59	0.26	0.81
		Cylinder	12.44	98.65	97.91	92.94
		Sphere	1.36	0.76	1.83	6.25
6		Box	58.07	1.68	20.81	0.35
		Cylinder	38.70	96.85	74.97	97.96
		Sphere	3.23	1.47	4.22	1.69
7	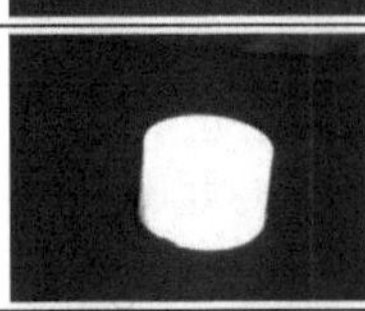	Box	2.74	2.42	0.63	9.01
		Cylinder	95.19	95.73	94.59	88.51
		Sphere	2.07	1.84	4.77	2.48
8		Box	0.48			
		Cylinder	25.46			
		Sphere	74.07			
9		Box	0.37			
		Cylinder	24.23			
		Sphere	75.40			

Table 5.3 Object classification percentages for different slants; test shapes

#	Object	Class	0°	30°	60°	90°
10		Box	98.83	85.80	85.07	85.55
		Cylinder	1.05	12.53	13.16	12.75
		Sphere	0.12	1.67	1.77	1.70
11		Box	94.60	89.97	85.07	91.08
		Cylinder	4.81	8.88	13.16	7.90
		Sphere	0.59	1.15	1.77	1.02
12		Box	80.49	96.20	86.02	91.72
		Cylinder	17.75	3.38	12.33	7.54
		Sphere	1.76	0.42	1.65	0.74
13		Box	94.48	95.59	90.59	99.43
		Cylinder	4.98	3.92	8.33	0.51
		Sphere	0.54	0.49	1.08	0.06
14		Box	0.59	5.86	0.33	0.51
		Cylinder	30.85	91.42	96.56	51.93
		Sphere	68.56	2.72	3.11	47.56
15	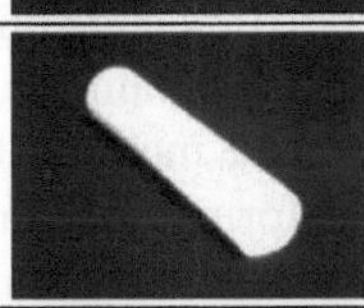	Box	60.35	61.66	35.89	0.32
		Cylinder	36.20	35.02	59.45	99.16
		Sphere	3.45	3.32	4.66	0.52
16		Box	57.89	84.91	93.33	98.05
		Cylinder	38.63	13.04	5.77	1.73
		Sphere	3.48	2.05	0.90	0.22
17		Box	0.82	1.03	0.42	0.71
		Cylinder	98.29	87.37	95.07	90.31
		Sphere	0.89	11.60	4.51	8.98
18		Box	17.89	0.20	3.69	3.81
		Cylinder	77.30	97.37	94.28	93.78
		Sphere	4.81	2.43	2.03	2.41

Even with this simplified setup, to reliably detect the salient points a double search is performed on the contour, combining the information given by different algorithms for corner (Teh and Chin 1989; Chetverikov and Szabo 1999) and edge detection (Ray and Ray 1995), to maximize the chance of finding all visible corners of the object when possible. A system able to segment the three faces of the object separately would provide a better estimate, but the results obtained with this simpler approach, presented below, match the application requirements.

Three variables identify object position and orientation. The distance d is measured between point P and the camera, and is enough to represent full 3D object location, as the object is centered in the image, and there is thus no lateral or vertical displacement. The other two variables are slant angle with respect to the frontoparallel position θ (see Fig. 5.4), and the direction of view with respect to the horizontal plane, ϕ. This last variable is known by the robot, and is computed by the vestibular system in primates. The viewing direction angle is restricted in the experiments, to allow a clear perspective view without simplifying too much the task as it happens for large angles (in such cases, the slant is very similar to what can be estimated simply using the inclination of segments in the 2D image). The final working range is about $15° < \phi < 50°$, and these are very plausible values even for a human subject looking at an object with grasping purposes. For what concerns the slant θ, only those situations that would reduce the interest of the slant estimation (for angles very close to 0° and 90°) are ruled out. These conditions can anyway be detected quite easily by the system, from the number and distribution of the defining corners.

The process of distance, pose and size estimation begins with the arm moving until point P of the object is placed horizontally at the center of the image, in order to minimize distortions due to the cameras' optics. Left and right images at this position are then processed: corners P, Q, R, W, T and U are found as explained above, and the position of S is estimated through a two point perspective method (Fig. 5.13). At this point, the coordinates of the defining points are transformed into angles with respect to the center of the image, using the camera focal lens and image size in pixels as parameters. The non-linearity of the camera optics is the reason to avoid getting close to the image borders, where distortions could affect the transformation process.

Once the six points identifying the two frontal faces of the object for both cameras have been detected, the actual slant estimation process can begin. Eight different estimators are calculated using the equations provided in Sect. 5.2.2: (5.8) is applied to the couples of segments PS/QR and UT/PS for both the left and right eye, whilst (5.6) is applied to segments PQ, SR, TS and UP. The first eight estimators, four perspective and four stereoscopic, of Table 5.4 are obtained at this point.

Before calculating the final, merged estimator it is useful to check for possible outliers (completely wrong estimations). In nature, bad estimations could be due to momentary occlusions, unusual light conditions, sudden movements, etc. In the simple setup used, any previous processing step can affect the final results, so again illumination issues, imperfections in the binarization or corner detection can cause one or more cues to deviate hugely from the average estimate. Outlier detection is a full sub-branch of statistics (Rousseeuw and Leroy 1987), and many different methods

Table 5.4 Slant estimators

#	Estimator	Computation method
1	Perspective I	Segments PS/QR, left eye
2	Perspective II	Segments PS/QR, right eye
3	Perspective III	Segments UT/PS, left eye
4	Perspective IV	Segments UT/PS, right eye
5	Stereopsis I	Segment PQ
6	Stereopsis II	Segment SR
7	Stereopsis III	Segment UP
8	Stereopsis IV	Segment TS
9	Merged ($\hat{\theta}_P$)	Perspective Only Average, # 1–4
10	Merged ($\hat{\theta}_S$)	Stereopsis Only Average, # 5–8
11	Merged ($\hat{\theta}_A$)	$\hat{\theta}_P$ and $\hat{\theta}_S$ Simple Average, # 9–10
12	Merged ($\hat{\theta}_G$)	Global Simple Average, # 1–8
13	Merged ($\hat{\theta}_W$)	Global Weighted Average, # 1–8

are available. Various techniques were explored, and they did not give significantly different results. The classical Rosner's many outliers test (Rosner 1975), widely used in the literature for similar problems, was finally chosen. The best results were obtained for a significance level $\alpha = 0.01$, which gave a final estimation improved of more than 5 % compared to the implementation without outlier rejection.

Following the model, monocular and binocular cues have to be merged according to their expected reliability and correlation. The starting point of the experiments is a situation in which no information is available regarding reliability of the different cues in the various working conditions. Therefore, to begin with, there are only two solutions readily available without the need of performing a training session for learning the cue weights. The first is to compute a simple, non-weighted average of a set of simple estimators (Estimators 9–12 of Table 5.4). The second is to compute an average in which weights are calculated using cue correlation (Estimators 13), in this case simply using the deviation of each cue from the simple average of all cues.

5.4.3.2 Nearness and Size Estimation

As no previous knowledge regarding the target object is assumed, it is not possible to disambiguate the pair distance/size only from retinal data. The nearness of the object can be calculated making use of expression (5.1), after estimating the proprioceptive vergence angle γ_P. The available stereo camera does not allow for vergence movements of the eyes, so they have to be simulated. The simple procedure adopted is to center point P of the object in one of the images first, and rotate the camera around the cyclopean eye, in order to center again P on the other image without changing the actual distance. To take advantage of this movement left and right images are taken both from the initial and the final position, and they are considered as two independent slant estimation experiments. No significant differences were observed regarding the estimation precision from the initial and the final position.

For what concerns size estimation, the relative size of the object (proportion between its edges) can be detected from orientation and separation angles alone. Once distances have been estimated, the actual dimensions of the object can be computed through simple geometric equations, as the ambiguity size/distance has been resolved.

5.4.4 Experimental Results

Overall, 422 experiments were executed with different values of slant and distance, as shown in Table 5.5. The global average estimation errors of all executed experiments are provided in Table 5.6. Perspective estimator $\hat{\theta}_P$ and stereopsis estimator $\hat{\theta}_S$ are calculated merging the four estimators of each modality alone. The simple average $\hat{\theta}_A$ is the mean between the two, and the global average $\hat{\theta}_G$ is the mean of all eight initial estimators. It is quite apparent how the combination of multiple cues, especially when they come from different kinds of visual information, strongly improves the estimation performance. The worst merged estimator $\hat{\theta}_P$ performs better than the best single cue estimator, Stereopsis I; the global average $\hat{\theta}_G$ improves the merged stereopsis estimator $\hat{\theta}_S$ by more than 25 %. The cue correlation weighted average estimator $\hat{\theta}_W$ shows a further improvement of around 8 % compared to $\hat{\theta}_G$, bringing the overall mean error close to 2.5°, which constitutes quite a good pose estimation for a robotic system, even in these restricted conditions.

Table 5.5 Number of experiments per distance and slant

Distance	Count	Slant	Count
450–500	14	10	12
500–550	40	20	80
550–600	66	30	96
600–650	74	40	80
650–700	88	50	92
700–750	94	60	48
750–800	28	70	14
800–850	18		
Total	422	Total	422

Table 5.6 Experimental slant estimation results, overall average errors

#	Estimator	Error(°)
1	Perspective I	8.63
2	Perspective II	6.67
3	Perspective III	12.75
4	Perspective IV	9.59
5	Stereopsis I	4.73
6	Stereopsis II	7.89
7	Stereopsis III	6.31
8	Stereopsis IV	5.41
9	Merged ($\hat{\theta}_P$)	4.71
10	Merged ($\hat{\theta}_S$)	3.92
11	Merged ($\hat{\theta}_A$)	3.78
12	Merged ($\hat{\theta}_G$)	2.91
13	Merged ($\hat{\theta}_W$)	2.68

5.4.4.1 Comparison with Human and Neural Network Simulation Data

It is interesting to compare error distributions obtained in the real practical experiments with the theoretical ones of Sect. 5.3.3. Figure 5.14 shows the average error plotted as a function of slant (Fig. 5.14a) and distance (Fig. 5.14b), similarly to Fig. 5.9. Some slant and distance values are probably affected by the use of different objects and viewpoints, which were not regularly distributed across conditions; see for example the bad quality of stereopsis, and consequently of the merged estimators, for slant =60°. Nevertheless, the trends are quite clear, and the expected effect of slant and distance on the different estimators is once again reproduced. In Fig. 5.14a the improvement in perspective estimation and the deterioration in stereoptic estimation with increasing slant are clearly visible, and the weighted average is definitely the best available estimator. Figure 5.14b shows that stereoptic estimation gradually decreases its precision with distance, whilst perspective seems nearly uncorrelated with it, apart for extreme values. Again the weighted average presents a clearly advantageous behavior in all cases.

It can be noted from both graphs how the weighted estimator maintains its reliability across conditions. Error bars of θ_W are always small apart from extreme conditions (which are also affected by a reduced number of trials). Errors for other estimators, which could not be plotted for clarity reasons, are always quite larger. This is a very important aspect for a robotic application, as there are no "blind spots" for which its estimation capabilities become unreliable. The implementation of a multiple cue estimation method thus provides a robotic system with a robustness hardly achievable with perspective or stereopsis alone.

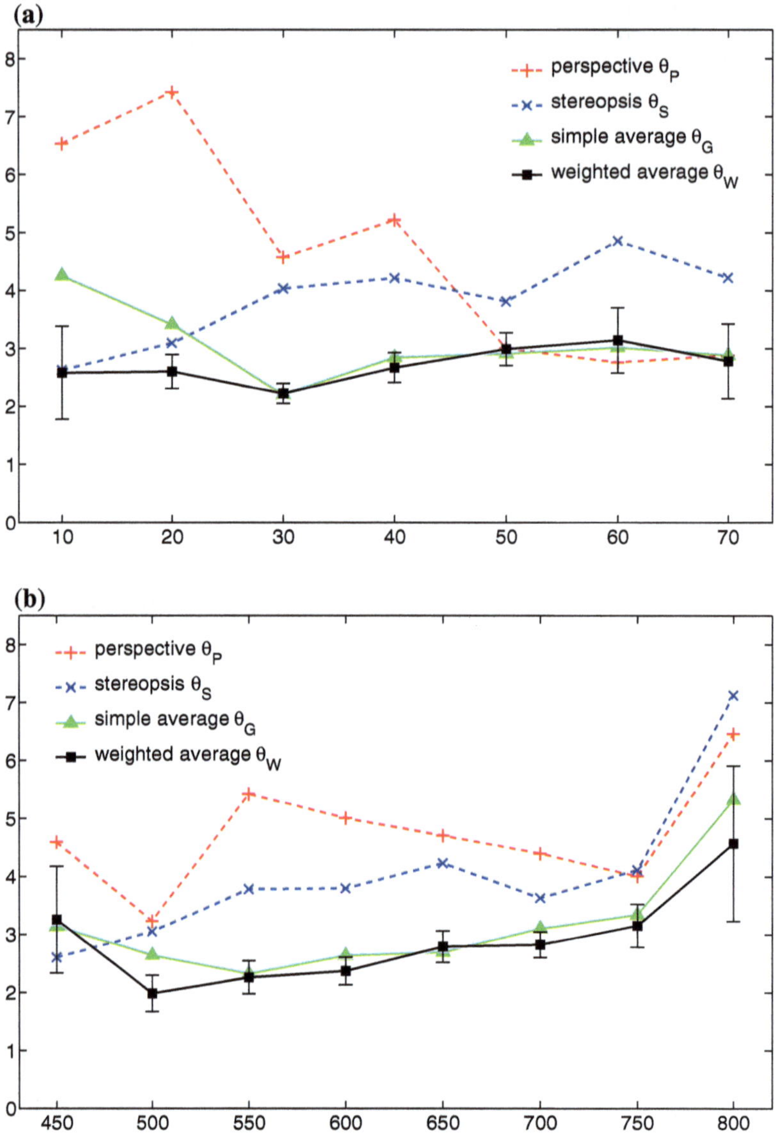

Fig. 5.14 Slant estimation error as a function of slant and distance; experimental results. For clarity, errors on errors are plotted only for θ_W. **a** Error (°) versus slant (°). **b** Error (°) versus distance (mm)

For what concerns distance estimation, the global average error for all experiments is of 33.4 mm, and the error distribution shown in Fig. 5.15, although noisy, follows the expected trend, showing decreasing estimation precision with increasing distance.

Size estimation revealed to be less precise compared to slant and distance estimation. In part, this is due to the fact that it makes use of two estimators and the

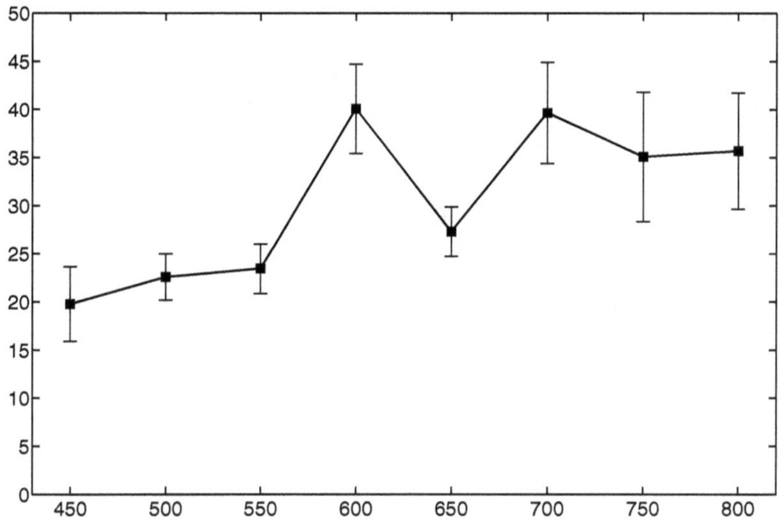

Fig. 5.15 Experimental distance estimation error; Error (mm) versus distance (mm)

theoretical final error is the product of the two initial errors. Moreover, for high slants and for small objects, the edges of the least visible side have very short separation angles, for which the relative error is much higher. Anyway, the worst case error is never larger than a few centimeters, and this is enough for reliable grasping by the robot hand, as shown in Sect. 6.2.4.

5.4.4.2 Additional Experiments

The second class of objects used to test the system were approximately cylindrical shapes, which still offer parallel edges. In the experimental setup, cylinders are lying on a plane, and the slant to estimate is that of their axis. Four salient points are detected on cylinder contours, those points in which curvature changes from 0 to some positive value, i.e., the transition from straight to curved segments. Those four points are treated as they were the P, Q, R and S of the box shapes (see Fig. 5.16). In this way, estimators 1, 2, 5 and 6 of Table 5.4 can be computed. The results of just few experiments are encouraging, as the average orientation estimation error is around 3.5°. It must be said though that the values chosen for viewpoint, slant and distance were those that gave the most consistent results in the experiments with boxes, and the method has not been tested exhaustively in various different conditions.

For spheres, only centroid, size and distance can be estimated. As for spheres there is no reliable point P with minimum y coordinate, the centroid was used instead in order to detect the vergence angle γ_P and hence the distance of the object. Distance computed with this method was found to have higher precision than for boxes and

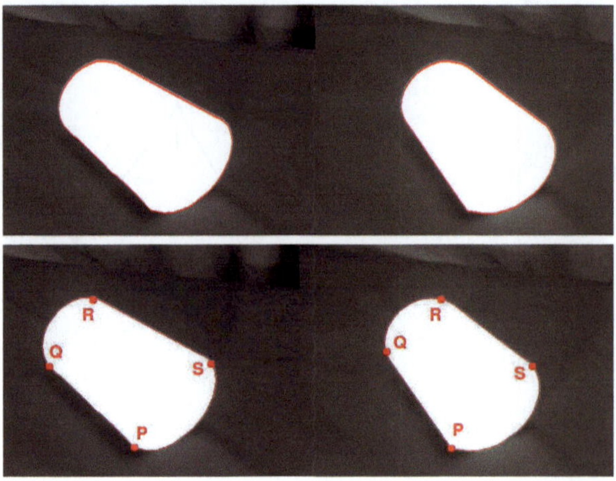

Fig. 5.16 Contour and salient points extraction for *cylindrical shapes*, *left* and *right* images

cylinders, and this reflects in an improved size estimation for spheres compared to the other classes.

5.5 Conclusions

The robotic implementation of the computational model for estimating object features in 3D permitted to achieve two important results. On the one hand, the robotic grasping system was provided with a very reliable and robust visual estimation of slant, distance and size of target objects. On the other hand, effects described in human experiments could be reproduced at a reasonable level of approximation. Cue integration is the fundamental principle which allowed to obtain such results, through the efficient merging of stereoscopic and perspective estimators.

The experimental results obtained with the robotic visual system confirm the hypothesis that integration of monocular and binocular data provide a robot with superior estimation capabilities. The final merged estimator obtained appropriately weighting the different cues is robust across working conditions, in a way that is probably not attainable by a simple estimator alone.

The following step is to make use of the extracted information regarding object potential grasping features to generate suitable action plans. This is the subject of next chapter.

References

Adams DL, Zeki S (2001) Functional organization of macaque V3 for stereoscopic depth. J Neurophysiol 86(5):2195–2203

Anzai A, Chowdhury SA, DeAngelis GC (2011) Coding of stereoscopic depth information in visual areas v3 and v3a. J Neurosci 31(28):10270–10282. doi:10.1523/JNEUROSCI.5956-10.2011

Backus BT, Banks MS, van Ee R, Crowell JA (1999) Horizontal and vertical disparity, eye position, and stereoscopic slant perception. Vision Res 39(6):1143–1170

Backus BT, Fleet DJ, Parker AJ, Heeger DJ (2001) Human cortical activity correlates with stereoscopic depth perception. J Neurophysiol 86(4):2054–2068

Backus BT, Banks MS (1999) Estimator reliability and distance scaling in stereoscopic slant perception. Perception 28(2):217–242

Banks MS, Hooge IT, Backus BT (2001) Perceiving slant about a horizontal axis from stereopsis. J Vis 1(2):55–79. 10:1167/1.2.1

Bar M, Tootell RB, Schacter DL, Greve DN, Fischl B, Mendola JD, Rosen BR, Dale AM (2001) Cortical mechanisms specific to explicit visual object recognition. Neuron 29(2):529–535

Blanz V, Tarr MJ, Bülthoff HH (1999) What object attributes determine canonical views? Perception 28(5):575–599

Bradshaw MF, Elliott KM, Watt SJ, Hibbard PB, Davies IRL, Simpson PJ (2004) Binocular cues and the control of prehension. Spat Vis 17(1–2):95–110

Brautaset RL, Jennings JAM (2005) Distance vergence adaptation is abnormal in subjects with convergence insufficiency. Ophthalmic Physiol Opt 25(3):211–214. doi:10.1111/j.1475-1313.2005.00274.x

Bray S, Arnold AEGF, Iaria G, MacQueen G (2013) Structural connectivity of visuotopic intraparietal sulcus. Neuroimage 82:137–145. doi:10.1016/j.neuroimage.2013.05.080

Brouwer GJ, van Ee R, Schwarzbach J (2005) Activation in visual cortex correlates with the awareness of stereoscopic depth. J Neurosci 25(45):10403–10413. doi:10.1523/JNEUROSCI.2408-05.2005

Bülthoff HH, Edelman SY, Tarr MJ (1995) How are three-dimensional objects represented in the brain? Cereb Cortex 5(3):247–260

Chetverikov D, Szabo Z (1999) A simple and efficient algorithm for detection of high curvature points in planar curves. In: Workshop of the austrian pattern recognition group, pp 175–184

Clark JJ, Yuille AL (1990) Data fusion for sensory information processing systems. Springer, Berlin

Cumming BG, DeAngelis GC (2001) The physiology of stereopsis. Annu Rev Neurosci 24:203–238. doi:10.1146/annurev.neuro.24.1.203

Deneve S, Pouget A (2003) Basis functions for object-centered representations. Neuron 37(2):347–359

Dobbins AC, Jeo RM, Fiser J, Allman JM (1998) Distance modulation of neural activity in the visual cortex. Science 281(5376):552–555

Einhäuser W, Hipp J, Eggert J, Körner E, König P (2005) Learning viewpoint invariant object representations using a temporal coherence principle. Biol Cybern 93(1):79–90. doi:10.1007/s00422-005-0585-8

Ekvall S, Hoffmann F, Kragic D (2003) Object recognition and pose estimation for robotic manipulation using color cooccurrence histograms. In: IEEE International conference on intelligent robots and systems, pp 1284–1289

Endo K, Haranaka Y, Shein WN, Adams DL, Kusunoki M, Sakata H (2000) Effects of different types of disparity cues on the response of axis-orientation selective cells in the monkey parietal cortex. Nippon Ganka Gakkai Zasshi 104(5):334–343

Ferrier N (1999) Determining surface orientation from fixated eye position and angular visual extent. In: IEEE international conference on robotics and automation, pp 938–943. doi:10.1109/ROBOT.1999.772426

Gehrig S, Badino H, Paysan P (2006) Accurate and model-free pose estimation of small objects for crash video analysis. In: British machine vision conference, Edinburgh

Genovesio A, Ferraina S (2004) Integration of retinal disparity and fixation-distance related signals toward an egocentric coding of distance in the posterior parietal cortex of primates. J Neurophysiol 91(6):2670–2684. doi:10.1152/jn.00712.2003
Georgieva S, Peeters R, Kolster H, Todd JT, Orban GA (2009) The processing of three-dimensional shape from disparity in the human brain. J Neurosci 29(3):727–742. doi:10.1523/JNEUROSCI.4753-08.2009
Germann M, Breitenstein MD, Park IK, Pfister H (2007) Automatic pose estimation for range images on the GPU. In: International conference on 3-d digital imaging and modeling, pp 81–90. doi:10.1109/3DIM.2007.13
Goddard J (1998) Pose and motion estimation using dual quaternion-based extended Kalman filtering. In: SPIE: three-dimensional image capture and applications, vol 3313
Gonzalez F, Perez R (1998) Neural mechanisms underlying stereoscopic vision. Prog Neurobiol 55(3):191–224
Greenwald HS, Knill DC, Saunders JA (2005) Integrating visual cues for motor control: a matter of time. Vis Res 45(15):1975–1989. doi:10.1016/j.visres.2005.01.025
Grill-Spector K, Kanwisher N (2005) Visual recognition: as soon as you know it is there, you know what it is. Psychol Sci 16(2):152–160. doi:10.1111/j.0956-7976.2005.00796.x
Grill-Spector K, Kushnir T, Hendler T, Edelman S, Itzchak Y, Malach R (1998) A sequence of object-processing stages revealed by fMRI in the human occipital lobe. Hum Brain Mapp 6(4):316–328
Heeley DW, Scott-Brown KC, Reid G, Maitland F (2003) Interocular orientation disparity and the stereoscopic perception of slanted surfaces. Spat Vis 16(2):183–207
Hillis JM, Watt SJ, Landy MS, Banks MS (2004) Slant from texture and disparity cues: optimal cue combination. J Vis 4(12):967–992. 10:1167/4.12.1
Howard IP, Rogers BJ (2002) Seeing in depth. I Porteous
Jacobs R (2002) What determines visual cue reliability? Trends in Cogn Sci 6(8):345–350
James KH, Humphrey GK, Goodale MA (2001) Manipulating and recognizing virtual objects: where the action is. Can J Exp Psychol 55(2):111–120
Jones DG, Malik J (1992) Determining three-dimensional shape from orientation and spatial frequency disparities. In: European conference on computer vision, pp 662–669
Julesz B (1971) Foundations of cyclopean perception. MIT Press, Cambridge
Knill DC (2007) Robust cue integration: a Bayesian model and evidence from cue-conflict studies with stereoscopic and figure cues to slant. J Vis 7(7):5.1-524. doi:10.1167/7.7.5
Konen CS, Mruczek REB, Montoya JL, Kastner S (2013) Functional organization of human posterior parietal cortex: grasping- and reaching-related activations relative to topographically organized cortex. J Neurophysiol 109(12):2897–2908. doi:10.1152/jn.00657.2012
Kragic D, Christensen HI (2001) Cue integration for visual servoing. IEEE J Rob Autom 17(1):18–27. doi:10.1109/70.917079
Landy MS, Maloney LT, Johnston EB, Young M (1995) Measurement and modeling of depth cue combination: in defense of weak fusion. Vision Res 35(3):389–412
Lehky SR, Pouget A, Sejnowski TJ (1990) Neural models of binocular depth perception. Cold Spring Harb Symp Quant Biol 55:765–777
Lehky SR, Sejnowski TJ (1990) Neural model of stereoacuity and depth interpolation based on a distributed representation of stereo disparity. J Neurosci 10(7):2281–2299
Lippiello V, Siciliano B, Villani L (2001) Position and orientation estimation based on Kalman filtering of stereo images. In: IEEE Int Conf Control Appl 702–707. doi:10.1109/CCA.2001.973950
Lippiello V, Siciliano B, Villani L (2006) 3D pose estimation for robotic applications based on a multi-camera hybrid visual system. In: IEEE International conference on robotics and automation, pp 2732–2737
Loftus A, Servos P, Goodale MA, Mendarozqueta N, Mon-Williams M (2004) When two eyes are better than one in prehension: monocular viewing and end-point variance. Exp Brain Res 158(3):317–327. doi:10.1007/s00221-004-1905-2

Marotta JJ, Perrot TS, Nicolle D, Servos P, Goodale MA (1995) Adapting to monocular vision: grasping with one eye. Exp Brain Res 104(1):107–114

Marr D (1982) Vision: a computational investigation into the human representation and processing of visual information. Freeman WH

Mon-Williams M, Tresilian JR, Roberts A (2000) Vergence provides veridical depth perception from horizontal retinal image disparities. Exp Brain Res 133(3):407–413

Mon-Williams M, Tresilian JR (1999) Some recent studies on the extraretinal contribution to distance perception. Perception 28(2):167–181

Naganuma T, Nose I, Inoue K, Takemoto A, Katsuyama N, Taira M (2005) Information processing of geometrical features of a surface based on binocular disparity cues: an fMRI study. Neurosci Res 51(2):147–155. doi:10.1016/j.neures.2004.10.009

Norman JF, Todd JT, Phillips F (1995) The perception of surface orientation from multiple sources of optical information. Percept Psychophys 57(5):629–636

Orban GA, Janssen P, Vogels R (2006) Extracting 3D structure from disparity. Trends Neurosci 29(8):466–473. doi:10.1016/j.tins.2006.06.012

O'Reilly RC, Munakata Y (2000) Computational explorations in cognitive neuroscience—understanding the mind by simulating the brain. MIT Press, Cambridge

Parker AJ (2004) From binocular disparity to the perception of stereoscopic depth. In: Chalupa LM, Werner JS (eds) The visual neurosciences, MIT Press, Cambridge, MA, chapter 49, pp 779–792

Perry CJ, Tahiri A, Fallah M (2014) Feature integration within and across visual streams occurs at different visual processing stages. J Vis 14(2). doi:10.1167/14.2.10

Peters G (2004) Efficient pose estimation using view-based object representations. Mach Vis Appl 16(1):59–63

Poggio GF, Gonzalez F, Krause F (1988) Stereoscopic mechanisms in monkey visual cortex: binocular correlation and disparity selectivity. J Neurosci 8(12):4531–4550

Pouget S, Sejnowski A (1997) Spatial transformations in the parietal cortex using basis functions. J Cogn Neurosci 9(2):222–237

Ray B, Ray K (1995) A new split-and-merge technique for polygonal-approximation of chain coded curves. Pattern Recogn Lett 16(2):161–169

Read J (2005) Early computational processing in binocular vision and depth perception. Prog Biophys Mol Biol 87(1):77–108. doi:10.1016/j.pbiomolbio.2004.06.005

Riesenhuber M, Poggio T (2000) Models of object recognition. Nat Neurosci 3:1199–1204. doi:10.1038/81479

Rolls ET, Webb TJ (2014) Finding and recognizing objects in natural scenes: complementary computations in the dorsal and ventral visual systems. Front Comput Neurosci 8:85. doi:10.3389/fncom.2014.00085

Rosenhahn B, Perwass C, Sommer G (2004) Pose estimation of 3D free-form contours. Int J Comput Vis 62:267–289. doi:10.1007/s11263-005-4883-3

Rosner B (1975) On the detection of many outliers. Technometrics 17(2):221–227

Rousseeuw PJ, Leroy AM (1987) Robust regression and outlier detection. Wiley, New York

Rutschmann RM, Greenlee MW (2004) Bold response in dorsal areas varies with relative disparity level. Neuroreport 15(4):615–619

Sakata H, Taira M, Kusunoki M, Murata A, Tanaka Y, Tsutsui K (1998) Neural coding of 3D features of objects for hand action in the parietal cortex of the monkey. Philos Trans R Soc B: Biol Sci 353(1373):1363–1373

Sakata H, Taira M, Kusunoki M, Murata A, Tsutsui K, Tanaka Y, Shein WN, Miyashita Y (1999) Neural representation of three-dimensional features of manipulation objects with stereopsis. Exp Brain Res 128(1–2):160–169

Salinas E, Thier P (2000) Gain modulation: a major computational principle of the central nervous system. Neuron 27(1):15–21

Saxena A, Schulte J, Ng AY (2007) Depth estimation using monocular and stereo cues. In: International joint conferences on artificial intelligence, pp 2197–2203

Shikata E, Hamzei F, Glauche V, Knab R, Dettmers C, Weiller C, Büchel C (2001) Surface orientation discrimination activates caudal and anterior intraparietal sulcus in humans: an event-related fMRI study. J Neurophysiol 85(3):1309–1314

Taira M, Tsutsui KI, Jiang M, Yara K, Sakata H (2000) Parietal neurons represent surface orientation from the gradient of binocular disparity. J Neurophysiol 83(5):3140–3146

Taira M, Nose I, Inoue K, Tsutsui K (2001) Cortical areas related to attention to 3D surface structures based on shading: an fMRI study. Neuroimage 14(5):959–966. doi:10.1006/nimg.2001.0895

Taylor G, Kleeman L (2003) Fusion of multimodal visual cues for model-based object tracking. In: Australasian conference on robotics and automation, Brisbane, Australia

Teh CH, Chin R (1989) On the detection of dominant points on digital curves. IEEE Trans Pattern Anal Mach Intell 11(8):859–872. doi:10.1109/34.31447

Thoma V, Henson RN (2011) Object representations in ventral and dorsal visual streams: fmri repetition effects depend on attention and part-whole configuration. Neuroimage 57(2):513–525. doi:10.1016/j.neuroimage.2011.04.035

Thomas OM, Cumming BG, Parker AJ (2002) A specialization for relative disparity in V2. Nat Neurosci 5(5):472–478. doi:10.1038/nn837

Todd JT (2004) The visual perception of 3D shape. Trends Cogn Sci 8(3):115–121. doi:10.1016/j.tics.2004.01.006

Tresilian JR, Mon-Williams M, Kelly BM (1999) Increasing confidence in vergence as a cue to distance. Proc R Soc B: Biol Sci 266(1414):39–44

Tresilian JR, Mon-Williams M (2000) Getting the measure of vergence weight in nearness perception. Exp Brain Res 132(3):362–368

Trotter Y, Celebrini S, Stricanne B, Thorpe S, Imbert M (1996) Neural processing of stereopsis as a function of viewing distance in primate visual cortical area V1. J Neurophysiol 76(5):2872–2885

Trotter Y, Celebrini S, Durand JB (2004) Evidence for implication of primate area V1 in neural 3-D spatial localization processing. J Physiol Paris 98(1–3):125–134. doi:10.1016/j.jphysparis.2004.03.004

Trucco E, Verri A (1998) Introductory techniques for 3-D computer vision. Prentice Hall

Tsao DY, Vanduffel W, Sasaki Y, Fize D, Knutsen TA, Mandeville JB, Wald LL, Dale AM, Rosen BR, Essen DCV, Livingstone MS, Orban GA, Tootell RBH (2003) Stereopsis activates V3A and caudal intraparietal areas in macaques and humans. Neuron 39(3):555–568

Tsutsui KI, Taira M, Sakata H (2005) Neural mechanisms of three-dimensional vision. Neurosci Res 51(3):221–229. doi:10.1016/j.neures.2004.11.006

Tsutsui K, Jiang M, Yara K, Sakata H, Taira M (2001) Integration of perspective and disparity cues in surface-orientation-selective neurons of area CIP. J Neurophysiol 86(6):2856–2867

Ullman S (1996) High-level vision. Object recognition and visual cognition. MIT Press, Cambridge

van Ee R, Banks MS, Backus BT (1999) An analysis of binocular slant contrast. Perception 28(9):1121–1145

von der Heydt R, Zhou H, Friedman HS (2000) Representation of stereoscopic edges in monkey visual cortex. Vision Res 40(15):1955–1967

Wandell BA (1995) Foundations of vision. Sinauer Associates

Watt SJ, Bradshaw MF (2003) The visual control of reaching and grasping: binocular disparity and motion parallax. J Exp Psychol Hum Percept Perform 29(2):404–415

Weigl A, Hohm K, Seitz M (1995) Processing sensor images for grasping disassembly objects with a parallel-jaw gripper. In: TELEMAN Telerobotics conference

Weiss K, Wörn H (2004) Tactile sensor system for an anthropomorphic robot hand. In: IEEE International conference on manipulation and grasping, Genova, Italy

Welchman AE, Deubelius A, Conrad V, Bülthoff HH, Kourtzi Z (2005) 3D shape perception from combined depth cues in human visual cortex. Nat Neurosci 8(6):820–827. doi:10.1038/nn1461

Chapter 6
Visuomotor Transformations for Grasp Planning and Execution

The previous chapter was dedicated to the processing of basic visual information aimed at extracting object properties, both spatial and cognitive, relevant for grasping purposes. This chapter deals with the tasks of transforming such properties into suitable hand configurations, and executing an appropriate grasping action on the target object. As a first step, the data extracted so far have to be expressed in a format especially dedicated to transformation into hand shapes. In Sect. 6.1, new computational descriptions are offered of the tasks performed by CIP as a fundamental relay station between the visual cortex and the visuomotor areas downstream. Analytical expressions of the transfer functions realized by surface and axis orientation selective neurons (SOS and AOS) of CIP are derived and discussed. Section 6.2 has a more practical stance, and describes how the obtained representations are used in grasp planning and execution. The different projections to AIP and its job as the fundamental hub in programming and monitoring grasping actions are discussed. Practical solutions are proposed for a working model of its connections with ventral stream, premotor cortex, somatosensory areas, basal ganglia and cerebellum (Fig. 6.1). Robotic grasping experiments based on such connections and exploiting tactile feedback for increased reliability are described.

6.1 Neural Coding in the Caudal Intraparietal Sulcus

The caudal intraparietal area, CIP, constitutes a central node in the spatial analysis processing of the dorsal stream, which endows the subject with the ability of interacting with her/his surrounding peripersonal environment. Neuroscience studies both on monkeys and humans have depicted a rather clear image of the sort of processing performed by CIP (see Sect. 2.3.1). At the computational level, though, this area has been rather neglected compared to its downstream neighbor AIP, more directly related to grasping actions.

As explained in Sect. 2.3.1, two main neuronal populations have been distinguished in CIP: surface orientation selective (SOS) and axis orientation selective

E. Chinellato and A.P. del Pobil, *The Visual Neuroscience of Robotic Grasping*, Cognitive Systems Monographs 28, DOI 10.1007/978-3-319-20303-4_6

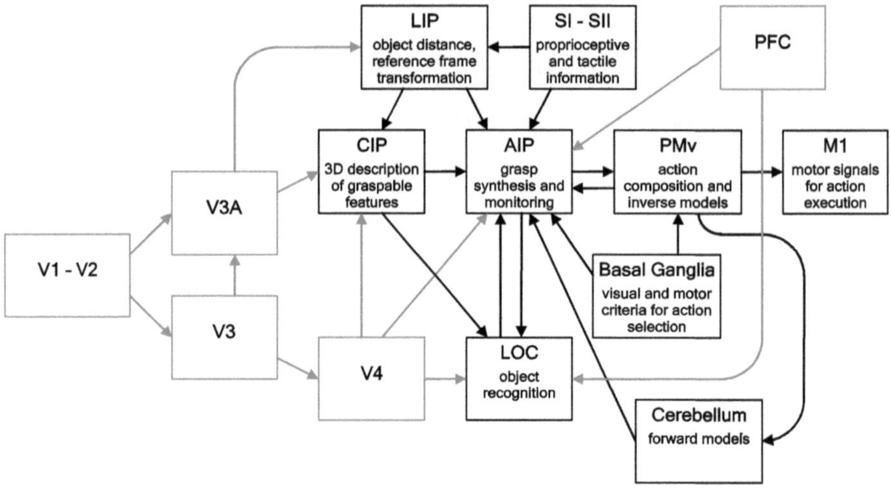

Fig. 6.1 Areas of the model framework involved in the planning and execution of grasping actions. The function of all highlighted areas is discussed in the text, but the implementation is especially focused on the job of AIP and its connections to the other areas

(AOS) neurons. SOS neurons code for the orientation of rather flat objects. Square shapes are preferred, and elongation in either width or length inhibits the neuronal response. The thickness of the object strongly inhibits the response only above a certain threshold. It can be hypothesized that such threshold represents the graspability of the feature, as it appears close to the size of the hand.

AOS neurons represent the 3D orientation of elongated objects, preferring thin and long features. It is not clear from the provided data if the reduced responsiveness to thicker objects is only due to the relative proportion between the object dimensions or also by a comparison with the hand size. The proposed model promotes this last possibility, for consistency with the role of CIP in providing AIP information regarding graspable features. Indeed, at least an approximate absolute object size estimation is available to CIP. Curvature coding, very likely also maintained in CIP, is not modeled at this point.

Overall, a population of mixed CIP neurons, including differently tuned SOS and AOS neurons, is able to provide full information about 3D proportion and orientation of a target shape. This information is forwarded to AIP, where 3D orientation, shape and size are jointly coded, and possible grip configurations generated.

Let us consider the situation in which a simple object (possibly box-like, or cylindrical) lies on a table, slanted about a vertical axis, as the ideal objects in Fig. 6.2. The goal is to generate, using only binocular visual information, possible grips on the object, emulating as much as possible the data flow connecting V3/V3A—CIP—AIP. In particular, the focus is on the tasks performed by the caudal intraparietal area, which can be schematized as in Fig. 6.3. The module on the left of the schema is the integration of proprioception with stereoscopic and perspective visual information

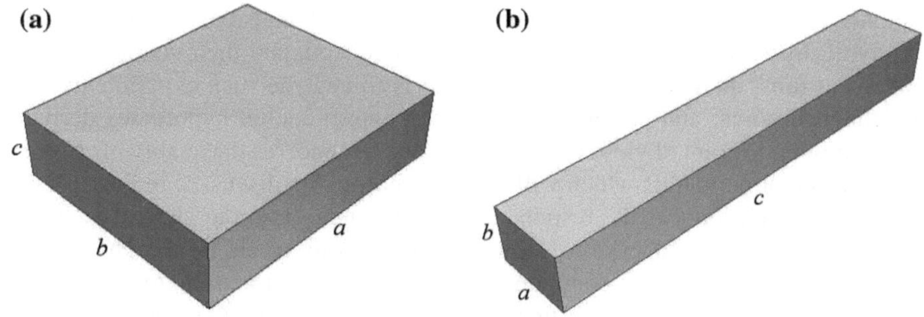

Fig. 6.2 Examples of SOS and AOS dominant objects and size naming convention. **a** Flat (SOS dominant) object. **b** Long (AOS dominant) object

in order to estimate position, orientation and size of simple 3D objects, introduced in the previous chapter.

The following step (right module of Fig. 6.3) requires an action-based point of view, to assess the intermediate level object features with the purpose of evaluating their suitability for grasping. Orientation, relative and absolute size of the major axes of the object are thus compared and the response is synthesized in the SOS and AOS neurons output. The activation of these two kinds of neurons depends on the relation between object dimensions. Considering the three main inertia axes, if two dimensions are similar and the third clearly smaller there is a high SOS activation; if two dimensions are similar and the third bigger AOS activation will prevail. In the case of three different dimensions, SOS and AOS responsiveness is modulated by the actual proportion between sizes. As a convention, from now on the three dimensions are called a, b and c, where a and b are close in size, whereas c is the smaller dimension for SOS activation (Fig. 6.2a) and the bigger dimension for the AOS case (Fig. 6.2b).

6.1.1 Understanding and Interpreting the Available Data

Despite the recent efforts and encouraging advancements (Naganuma et al. 2005), the most important insights regarding the nature of 3D object representation by CIP neurons date back to the second half of the last decade (Shikata et al. 1996; Sakata et al. 1998). The basic concepts were clear from the beginning, such as the distinction between the two classes of orientation responsive neurons, SOS and AOS, and their

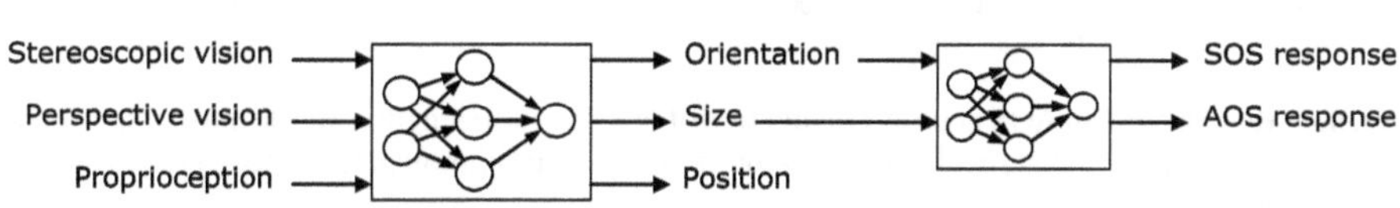

Fig. 6.3 Elaboration of visual data in the posterior intraparietal sulcus CIP

responsiveness trend as a function of an object relative dimensions. The number and variety of different experiments is nevertheless reduced, and their characterization remains at most qualitative. The current goal is to analyze such experiments with modeling purposes, and possibly advance new interpretation hypotheses deriving from a pragmatic point of view. One such hypothesis concerns the quality of absolute size representation in CIP, which will be more thoroughly discussed in Sect. 6.2.1.

Figure 6.4a reproduces the response of an AOS neuron to the view of a slanted elongated object as a function of object width (Sakata et al. 1998). The authors of the original study briefly comment on it suggesting that neuronal response and object width are inversely proportional. A sigmoidal, or logistic response function constitutes an alternative explanation. This solution fits very well with the observed data, as can be observed in Fig. 6.4b, where two differently parameterized sigmoidals are superposed to the data of Fig. 6.4a. The sigmoidal is a transfer function very commonly found in brain mechanisms (see e.g. Hu et al. 2005), especially when some threshold effects have to be taken into account. Indeed, in this case there is a very important threshold to consider, that is, the size of the grasping hand. The assumption is that the cut-off value for the sigmoid is the dimension of the open hand or even better the extension of a comfortable grip. For the monkey performing the experiment of Fig. 6.4a this value is reasonably around 12–15 cm. Indeed, CIP neurons seem to be sensitive not only to relative object dimensions (and thus shape) but also to its absolute size (Sakata et al. 1997). Available experimental data are not conclusive to this regard though. In any case, if the size of a potentially graspable object has to be represented in the brain, hand size is a very useful and convenient unit of measure to use. In the next section, this principle is further developed and exploited for defining the analytic expressions which model the function of SOS and AOS neurons.

Overall, CIP is responsive for all the following features of an object: relative size of main axes, absolute size, orientation in 3D, local curvature. Studies reported

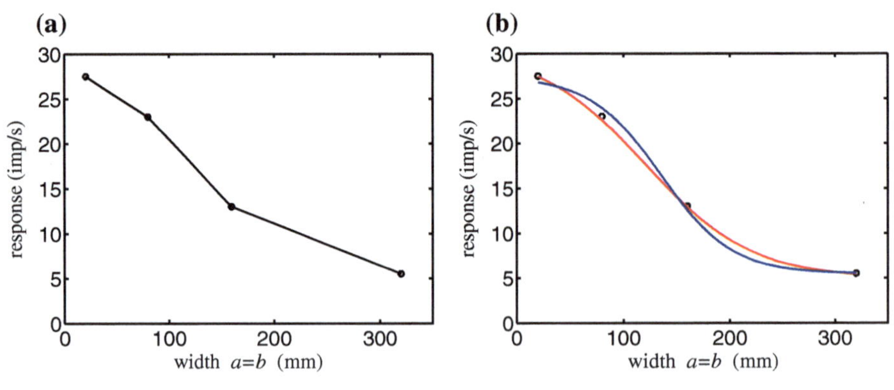

Fig. 6.4 Response of an AOS neuron as a function of object width (length $c = 300$ mm). Experimental data (adapted from Sakata et al. 1998) and interpolation with sigmoidal functions. **a** Experimental data. **b** Sigmoidal interpolation

in the literature describe SOS and AOS neurons that are selective only for width and not for thickness, or only for relative and not for absolute size. Indeed, just a minority of CIP neurons are selective for all the features at the same time, but globally, at a population level, all relevant information regarding object shape in relation to potential grasping actions is processed by the posterior intraparietal area (Sakata et al. 2005). The proposed transfer functions take into account dimensional aspects at a neural population level, leaving aside orientation extraction (modeled in Chap. 5) and curvature.

6.1.2 SOS Neurons Transfer Function

As a general principle, SOS neurons are preferentially activated when two dimensions of the object are similar, while the third is sensibly smaller: $a \geq b \gg c$. Experiments performed varying the width and the thickness of the object gave the results reproduced in Fig. 6.5 (Shikata et al. 1996). These graphs and the comments of the authors, together with the principles previously introduced, are the bases for defining a transfer function which models the behavior of a population of SOS neurons.

The proposed transfer function depends on three main factors represented by three penalty, or inhibition terms, that take into account different aspects of SOS neurons responsiveness. In a hypothetical ideal situation, all inhibition terms would be zero and activation maximal.

The first component of the transfer function is I_s, the *symmetry inhibition term*. This term takes into account the difference between the two major dimensions of the object a and b: responsiveness is maximal, and inhibition minimal, for equal major axes. Asymmetrical situations are given higher penalties. The value of I_s is 0 when the major dimensions are equal, and increases with their difference:

$$I_s = \left(\frac{a-b}{a+b}\right)^{k_s} \tag{6.1}$$

Constant k_s modulates the effect of the difference between a and b on I_s. The exact value of k_s can be deduced only experimentally, and is not necessarily stable across conditions.

The second term considers the relation between the minor, most easily graspable dimension, c and the major ones a and b. It is called I_f, *flatness inhibition term*, and it increases with dimension c, representing the thickness of the object:

$$I_f = \frac{c}{a+b} \tag{6.2}$$

The two previous terms are independent from the absolute size of the object. As discussed in the previous section, it is though likely that the hand size is playing an important role in determining the global responsiveness of CIP to a given target object. The *graspability inhibition term* I_g was thus introduced. As anticipated, it is expressed as a sigmoidal function. I_g decreases when increasing the graspable dimension c, and its symmetry point is the limit of a comfortable hand opening, called H:

$$I_g = \sigma(c, H) = \frac{1}{1 + e^{-k_g(c-H)}} \tag{6.3}$$

Constant k_g affects in this case the non-linearity of the equation: the larger k_g, the steepest the slope of the sigmoid function, and thus the influence of hand size H on SOS activation.

The global response R_{SOS} of a population of SOS neurons is thus estimated detracting the inhibitory quantities, appropriately weighted, from the theoretical 100 % activation:

$$R_{SOS} = 1 - w_s \cdot I_s - w_f \cdot I_f - w_g \cdot I_g \tag{6.4}$$

The given expression is still undetermined, as the two parameters k_s and k_g and the three weights w have not been assigned any value yet. Starting with the symmetry term alone, least squares fitting can be used to compute the value of k_s and w_s that best fits (6.1) to the data corresponding to Fig. 6.5a. This gives $k_s = 1.948$, and $w_s = 1.059$. It looks reasonable to simplify setting $k_s = 2$ and $w_s = 1$. In this way, I_s is the square of the fraction $(a - b)/(a + b)$ and its weight can be omitted. Similarly, (6.3) can be fitted to the data of Fig. 6.5b. A value of 0.042 is obtained for the estimation of k_g, and 0.458 for w_g. With a little approximation, $k_g = 0.04$ and $w_g = 0.5$. Finally, the only remaining coefficient w_f is estimated through least squares fitting of (6.2) to the data of Fig. 6.5b (taking into account the contribution of (6.3)). The final result is $w_f = 0.030$. After substituting all these values in the corresponding formulas, the response of (6.4) remains more explicitly defined as:

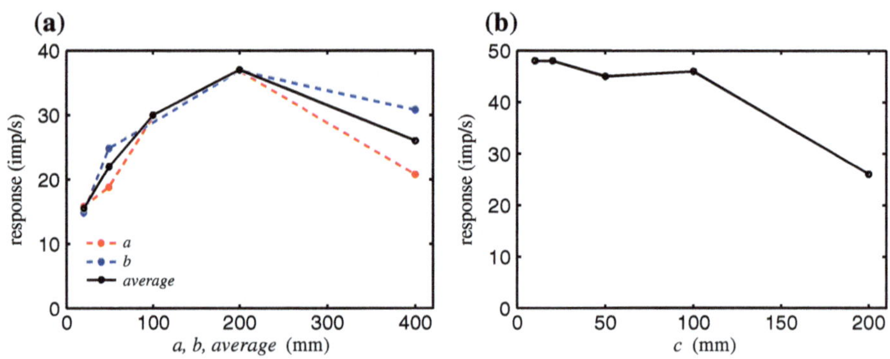

Fig. 6.5 Response of an SOS neuron as a function of object width and thickness. For width, the two major size responses and their average are plotted; the constant major size is 200 mm. Experimental data adapted from Shikata et al. (1996). **a** Width response (thickness $c = 20$ mm). **b** Thickness response (width $a = b = 200$ mm)

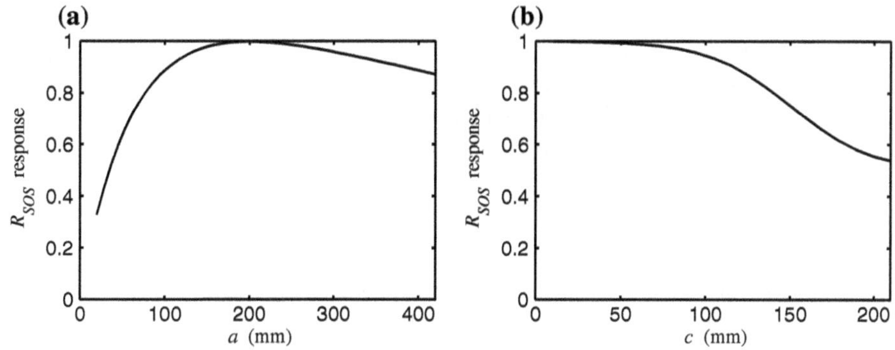

Fig. 6.6 Response of an SOS neuron as a function of object width and thickness. Simulated data obtained with (6.5), for the same width and thickness values of Fig. 6.5. **a** Width response ($b = 200$ mm, $c = 20$ mm). **b** Thickness response ($a = b = 200$ mm)

$$R_{SOS} = 1 - \left(\frac{a-b}{a+b}\right)^2 - 0.03\frac{c}{a+b} - 0.5\frac{1}{1+e^{-0.04(c-H)}} \qquad (6.5)$$

The global SOS response according to 6.5 was calculated as a function of object width and thickness. The results depicted in Fig. 6.6 show how the proposed model, properly parameterized, nicely fits the experimental data of Fig. 6.5 ($H = 150$ mm).

6.1.3 AOS Neurons Transfer Function

Axis orientation selective (AOS) neurons activate when one of the three dimensions of the object is quite larger than the other two, which are closer in size: $c \gg a \geq b$. Compared to SOS, less numerical results are available in the literature, and the main source of information is Fig. 6.4a, with the description of the corresponding experiments (Sakata et al. 1998). SOS and AOS neurons are intermixed in CIP, and it is thus plausible to assume that their response functions are similar. The hypothetical transfer function of AOS neurons was thus composed starting from the same three inhibition terms introduced in the previous section.

AOS *symmetry inhibition term* is equal to 0 when dimensions a and b are equal, and increases proportionally with their difference, exactly as in (6.1):

$$I_s = \left(\frac{a-b}{a+b}\right)^{k_s} \qquad (6.6)$$

No experiments explicitly designed to verify the effect of differences between the two minor dimensions have been carried out for AOS neurons. This effect is probably not very strong, but it can be reasonably assumed that a large asymmetry would indeed affect the perception of the elongated object. Such reduced influence of the fraction $(a-b)/(a+b)$ on the total response can be obtained changing the constant k_s.

Similarly to (6.2), the next term compares the major and minor dimensions of the object. This time, it is called I_l, *length inhibition term*, as it decreases with increasing the major dimension c of the object:

$$I_l = \frac{a}{c} \tag{6.7}$$

The graspable dimension, a in this case, is again the numerator of the fraction, as was c in (6.2). In this case the numerator could also be $(a+b)/2$, but if a and b are very similar this would likely be a pointless calculation.

The *graspability inhibition term* is again a sigmoidal function decreasing with the increasing of the minor dimension a, having as symmetry point the limit of a comfortable hand opening H.

$$I_g = \sigma(a, H) = \frac{1}{1 + e^{-k_g(a-H)}} \tag{6.8}$$

Again, the activation of a population of AOS neurons is estimated detracting the inhibition quantities from the theoretical 100 % activation:

$$R_{AOS} = 1 - w_s \cdot I_s - w_l \cdot I_l - w_g \cdot I_g \tag{6.9}$$

Due to the limited availability of data, a bigger extrapolation effort is needed in the AOS case to estimate appropriate values for parameters and coefficients. The case of the symmetry term is the most critical, as there is no published numerical data which can help in determining the values of k_s and w_s. This second coefficient can be set to the same value as for SOS neurons, $w_s = 1$, whilst k_s should be assigned a value such that the influence of the term on the overall response is reduced with respect to the SOS case. The easiest solution, but certainly no the only possible one, is to set $k_s = 1$, and leave only the fraction component. Response would thus linearly increase when reducing the difference between a and b. Regarding graspability, there are no reasons to believe that parameter k_g and weight w_g should be much different from the SOS case. Least squares fitting of (6.8) to the data of Fig. 6.4a gives values included in [0.02, 0.05] for k_g and in [0.5, 0.8] for w_g, depending on the initial conditions. It seems thus reasonable, for symmetry and ecological reasons, to set $k_g = 0.04$ and $w_g = 0.5$, as in (6.5).

Sakata et al. (1998) state that: "discharge rate of the AOS neurons increased monotonically with increasing length of the stimulus". The authors did not provide further information on this issue, but this comment describes how to generate additional data which could help in fitting the functions. A small additional dataset of 6 points in which response linearly increases with c was thus prepared. The newly generated dataset was used to fit (6.7) and thus set the value of w_l. Values between 0.2 and 1 were obtained using different graspable sizes of a. There is no reason why

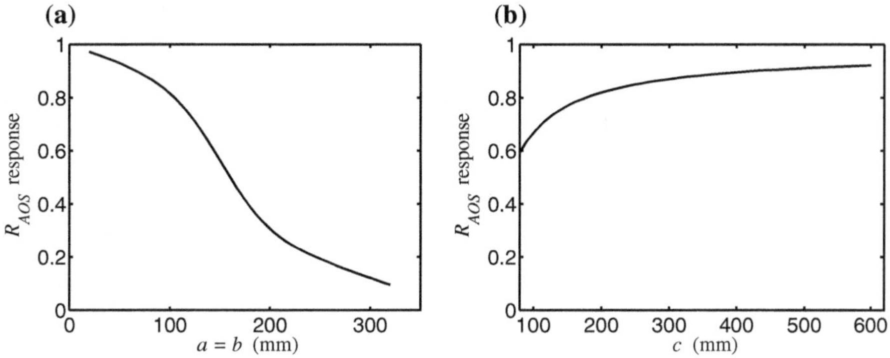

Fig. 6.7 Response of an AOS neurons as a function of object width and length. Simulated data obtained with (6.10). **a** Width response ($c = 300$ mm). **b** Length response ($a = b = 80$ mm)

the value of w_l should not change dynamically, but for the moment an intermediate value of $w_l = 0.374$, obtained for $a = 80$, is chosen. The overall formula for AOS response is thus defined as:

$$R_{AOS} = 1 - \frac{a-b}{a+b} - 0.37\frac{a}{c} - 0.5\frac{1}{1+e^{-0.04(a-H)}} \tag{6.10}$$

The behavior of (6.10) with changing thickness and length of the object is shown in Fig. 6.7 ($H = 150$ mm). Figure 6.7a tries to reproduce the effect depicted in Fig. 6.4a, whilst Fig. 6.7b shows how the response grows when increasing c. Again, the effects described in the neuroscience literature are well reproduced.

6.1.4 Robotic SOS and AOS

After definition of the transfer functions and comparison with the available neuroscience data, the CIP neuron model can be tested on the robotic setup with images of real objects. Using the object pose estimation procedure described in Chap. 5, the dimensions of twelve shapes, depicted in Fig. 6.8, were estimated and used to compute simulated AOS and SOS activations for each shape. The shapes used for this purpose were eight boxes and four cylinders of different size and proportions. In principle, the modeled activations do not take into account curvature and do not distinguish between cylinders and parallelepipeds. Nevertheless, all cylinders have the same a and b dimensions (their diameter), and this increases their AOS responsiveness, because the length inhibition term (6.7) is always 0.

The constants of the final activation functions were employed, with the exception of k_s in (6.5), which was set to 1 as in (6.10), in order to improve the equilibrium between SOS and AOS activations. Average activation across 10 trials for all the

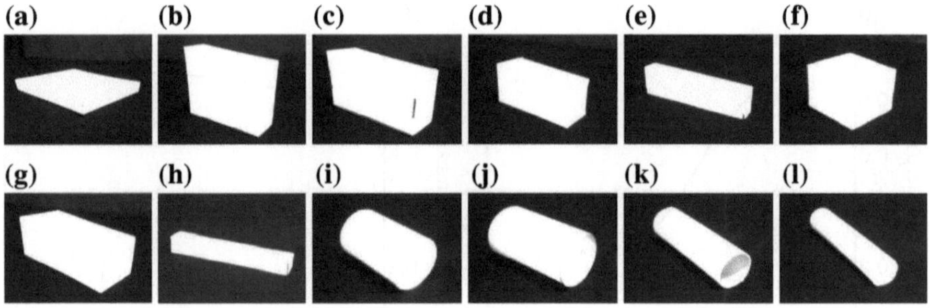

Fig. 6.8 Shapes for which experimental SOS and AOS activations are computed

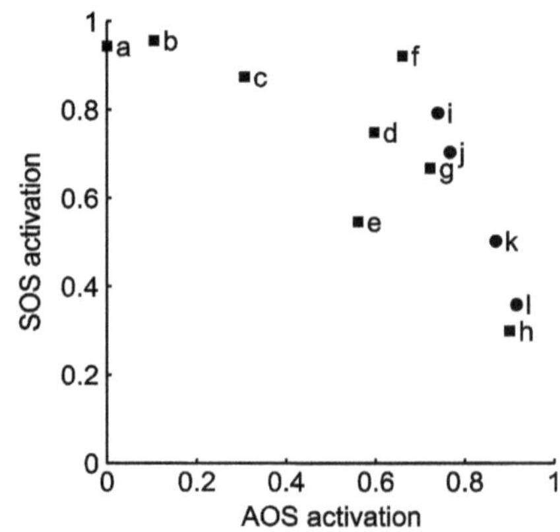

Fig. 6.9 Experimental SOS/AOS activation for the shapes of Fig. 6.8

shapes in Fig. 6.8 is mapped on an SOS/AOS graph displayed in Fig. 6.9. Standard deviations are very low and hence not plotted. The comfortable grasp size H was maintained equal to 150 mm: although the robot can grasp larger objects, 150 mm is about the limit that allows for full contact of the tactile sensors placed on the robot hand fingertips.

Even though there is no direct comparison available for validating the obtained results, a visual assessment of the activations plotted in Fig. 6.9 reveals that activations look appropriate for the objects' characteristics. The only clearly elongated box, (h), shows a clear dominance of AOS over SOS response. On the other extreme, boxes (a) and (b) are undoubtedly assessed as completely flat. For box (c) the SOS activation is still clearly superior to the AOS activation, whilst for (d) and (f) the difference is much reduced. Boxes (d), (e) and (g) demonstrate a substantial equilibrium between activations, with a light bias toward SOS for (d) and toward AOS for (g). It is interesting to observe that nearly all boxes are disposed along an arc from (a) to (h), with only (e) and (f) deviating from the main path. Such deviations can be simply a side-effect of the model approximation, but could also reflect an increased suitability for grasping actions of (f) with respect to (e). For what concerns cylinders, (i) is the only one having a larger SOS activation, while for (j) it is the AOS responsiveness that is slightly prevailing. Cylinders (k) and (l) are clearly elongated, and their AOS activation is dominant. Qualitatively, these results seem to properly represent the range of possible object proportions. From a robotic point of view, they show that the system is able to properly detect and code absolute and relative dimensions of target objects. For the model, these results suggest that it goes in the right direction, but more neuroscience experiments of different kinds would be needed for refinement and further validation.

Anticipating the possible use of SOS and AOS activations for the generation of hand configurations, the analyzed objects can be easily clustered in three groups. Objects on the top left of the graph, (a), (b) and (c), are definitely flat and will likely be grasped with a pad opposition between thumb and fingers. Elongated objects (k), (l) and (h) form the bottom right cluster, denoting AOS dominance, and can be grasped with either a precision grip or an involving grip. More complicated is the situation for the six objects in the central cluster. In fact, for hand shaping, they seem to be more different one another than represented in the graph. Size and curvature are probably the factors that would further distinguish between them to drive the selection of suitable grips.

6.1.5 Discussion and Future Developments

The above model offers some solutions to the problem of identifying the transfer functions of the different areas of the dorsal stream, but opens at least as many questions. More experiments are needed to validate the proposal. The actual importance of hand size on SOS and AOS activation should be explicitly analyzed, through experimental protocols designed to distinguish the effect of relative and absolute size of features. For example, no experiments are reported in the CIP literature regarding non graspable (or strangely shaped) objects, and these are definitely required at this point. Similarly, there is the need to disambiguate the influence of shape and size on neuronal response. This can be done by gradually changing the proportion and size of objects, and analyzing the response as a function of only one driving variable at a time. The responsiveness to object curvature should also be further explored. As the robotic simulation pointed out, it is very likely that the proposed functions will need to be updated and suited to new findings and requirements, but they constitute a helpful tool for orienting the future studies on the subject. The next step in the grasp planning process, performed by AIP, is to join SOS and AOS activations with data coming from other brain areas for deciding how to grasp possible target objects.

6.2 Planning and Executing the Grasping Action

The coding of SOS and AOS neurons for the visual characteristics of objects relevant for grasping purposes is the ideal input for AIP so that it can process the visual data and transform it to suitable hand configurations. A critical question is how much of the information that AIP needs is provided by CIP and how much needs to be complemented by other areas.

6.2.1 Characteristics of the Visual Input to AIP

Fukuda et al. (2000) accurately measured, with a data glove, human grasping configurations on fifteen different objects of three classes (spheres, cylinders and parallelepipeds) and five different sizes for each class. They registered the values of 18 joint angles of the hand at the time of contact in real and *pantomimed grasping* (see related text box), extracting the two principal components of the joint space for each condition. They found statistically significant differences between real and pantomimed grasping. To a minor extent, they also found a difference between real and 2D object stimuli in pantomimed grasping. Both findings are consistent with the two streams literature. They also demonstrated with a neural network implementation that the visual information provided to the subject could account for 99 % of the variability observed in joint configurations, suggesting that indeed grasping was purely based on the available visual data. These results would have probably been very different if task and object identity were taken into account.

A comparison of the modeled SOS/AOS representation with the work of Fukuda et al. (2000) can help in drawing some conclusions on the completeness of the proposed model. In Fig. 6.10 an adaptation of the results provided in the cited work is compared with a representation of the same objects in a simulated SOS/AOS activation space. Figure 6.10a shows an average, across the three experimental subjects, of the first and second principal components of the joint space when grasping spheres of different sizes and cylinders and parallelepipeds of different thicknesses. Objects

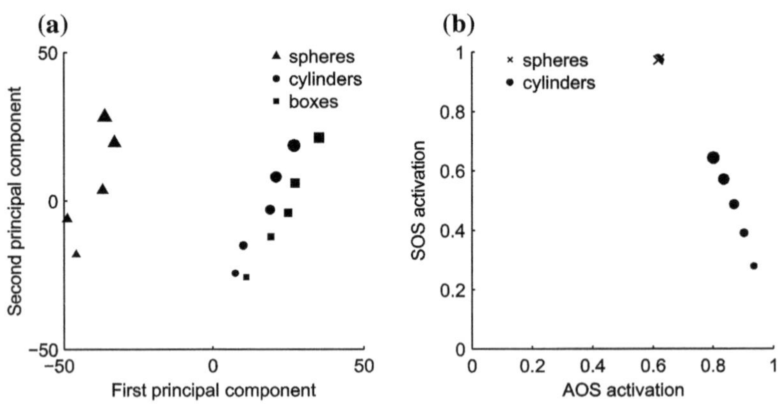

Fig. 6.10 Comparison between principal components of joint space during grasping (adapted from Fukuda et al. 2000) and AOS/SOS coding of similar objects. **a** Grasping joint space. **b** AOS/SOS representation

of the same dimensions were used to calculate the AOS and SOS activations of Fig. 6.10b.

The model does not recognize curvature, so parallelepipeds and cylinders with similar proportions have similar representations, and the same happens for cubes and spheres. The comparison between joint spaces for grasping cylinders and parallelepipeds (Fig. 6.10a), which are very similar, partially justifies this simplification. Whereas Fig. 6.10a represents an output of AIP processing, Fig. 6.10b constitutes a possible, probably partial, input. A qualitative comparison suggests that, while for cylinders the visual information provided by AOS and SOS neurons seems to be enough for generating an appropriate joint space for grasping, spheres of different sizes show nearly the same representation, suggesting that there is not enough data for deciding on how to grasp them. The reason for this discrepancy is probably twofold. First, AOS and SOS activation models were built on the existing data, which do not take into account all aspects of shape estimation. Size effects in the model are clearly observable only for the biggest shapes, close to the hand opening threshold. Indeed, activations of CIP neurons for equal objects of different sizes are not provided in the literature, and such aspect is not fully taken into account.

The second reason is related to the tasks and connectivity of AIP. Distance estimation, performed in LIP, is critical for the reliability of objects' size estimation. Ventral stream information regarding recognized objects also carries very good size estimates derived from experience. Some CIP neurons also code for distance (Shikata et al. 1996; Sakata et al. 1998), but it is likely that projections from LIP and from the ventral stream provide AIP with more exact estimates of the object size. It is thus likely that CIP uses a distance, and hence a size estimate, less precise than the one available to AIP. A possible hypothesis is that object size representation in CIP is exploited only with the purpose of filtering graspable from non graspable features, leaving exact estimation to LIP and AIP. The coding of potentially graspable object features, conveyed by the firing of SOS and AOS neurons, needs hence to be completed by accurate data on object size and location.

In spite of its critical importance, visual information is only one ingredient of the complex grasping recipe. In the next section, additional important factors, such as the criteria to follow in order to obtain reliable grasping actions, are described.

6.2.2 The Search for Grasp Quality

After AIP has gathered the available visual data regarding a target object, a number of issues have to be taken into account in order to produce an appropriate grasp. A very important aspect that strongly affects grasp planning is the search for quality in a grasping action. The same definition of quality is controversial, as strictly related to the task to execute. If the task is to handle a pen for writing, quality is measured in terms of manipulability. If the goal is to lift a heavy object, stability in the grip has to be pursued. For the limited scope of robotic grasping, often quality has been interpreted as a synonym of stability, or reliability of the action. The reduced tactile skills of robots compared to primates makes of reliability a critical factor for the selection of a grip. Especially if no or little experience is available, deciding among candidate affordances represents a key task (Borst et al. 2004; Morales et al. 2004). Human beings take into account a number of aspects which help in ensuring at least a minimum level of reliability to their grasping actions, as described in Sect. 4.3.3.1.

Multiple criteria and measures have been employed for assessing and predicting robot grasp quality (Roa and Surez 2015). In Chinellato et al. (2005), a set of visual criteria for the reliability assessment of planar grips was defined, taking as reference physiological studies of human grasping, and robotic research on grasp stability. The criteria were used to predict the outcome of future robot grasping actions (Chinellato et al. 2003; Morales et al. 2004). Some of those criteria can be extended, maintaining their plausibility and usefulness, to the three-dimensional case, and are presented below. An important conceptual difference is that in Chinellato et al. (2005) criteria were computed for a number of pre-defined candidate grips, in order to globally evaluate them and select one of them for execution, whilst here the task is to generate a grasp plan that implicitly achieves good quality values. The criteria are consistent with the findings described in Sect. 4.3.3.1, and it can be hypothesized that there is an important contribution of the basal ganglia in providing AIP with the signals required to select between alternative grasping patterns (Clower et al. 2005). In fact, it has been suggested that the basal ganglia is a key area in the development of action selection tasks through reinforcement learning (Doya 1999).

The quality criteria can be subdivided into two classes: visual criteria, that mostly affect the selection of the contact points on the object; and motor criteria, that mostly affect hand shaping.

6.2.2.1 Visual Criteria

At least three visual criteria important in human grasping are also useful for robotic implementation.

Center of mass. The opposition axis of the grip should always pass close to the object center of mass, in order to minimize the effect of gravitational and inertial torques, especially if the object is heavy. If possible, grasping along the main inertia axes is preferred for the same reasons. Moreover, heavy objects are often grasped above the center of mass for increased stability (Bingham and Muchisky 1993). In many cases, this criterion is predominant in human grasping.

Grasping margin. This criterion aims at minimizing the risk of placing the fingers on unsuitable object features which could result in unstable contacts. It builds on the assumption that fingers should be placed far from edges, and that large grasping surfaces, at least above a given threshold, should be chosen if available.

Curvature. Grips on slightly concave surfaces are normally considered more reliable than grips on convex ones, because contact surface and thus friction is higher in the first case (Jenmalm et al. 2000). To implement this criterion computationally, the curvature of graspable features could be calculated at different frequencies, as a slowly changing curvature is normally preferred. On the other hand, very high frequency curvature changes may indicate the presence of a rough surface, which is good to grasp because it offers high contact friction.

6.2.2.2 Motor Criteria

Finger extension This criterion aims at maximizing the contact surface between fingertips and object. The goal is to have a substantial equilibrium between the opening of all fingers. Moreover, an average finger aperture is preferred as it allows for bigger contact surfaces. Even though for humans this aspect is less important, because of the number of degrees of freedom of the hand and the high compliance of the fingertips, a grasping action in which some fingers are extended and other flexed is usually clumsy.

Force distribution In many cases, the optimization of the previous criterion ends in an optimal distribution of forces as well. Nevertheless, in case of complex objects and grasps with abducted fingers, a homogeneous distribution of the contact forces can be a critical issue. Usually, badly equilibrated grips can be improved through tactile feedback, but if the force asymmetry is too high, the object could slip or rotate due to unwanted torques.

An estimation of the movement cost should be added to these criteria to take into account the reaching comfort (see Sect. 4.3.3.1). This can be done by computing the expected joint rotations required to achieve a given goal position.

6.2.2.3 Modulation of the Effect of Quality Criteria

The ventral stream contribution provides the use of the quality criteria with an important flexibility. Knowledge regarding object characteristics, such as weight or compliance, and the outcome of previous grasping experiences can be used as modulation factors which assign different importance to the above criteria in different conditions. Default, prudential solutions are adopted in the case of failed recognition or low classification confidence, to respect the uncertainty of the situation. If the object properties are known, the biasing toward one criteria or another can be much stronger. To give a simple example, if the object is big and heavy, the center of mass criterion and the force distribution are very important, whilst for a small light object the grasping margin is probably the critical criterion. Recent psychophysiological findings support this hypothesis (Eastough and Edwards 2007). Computationally, criteria weighting can be initially hard-wired, but when the system increases its knowledge of the graspable world, this aspect should acquire a more dynamical behavior, especially if feedback is available regarding the appropriateness of grasping decisions.

6.2.3 Grasp Planning

All elements necessary to generate a grasping action suitable to a given condition have been described, and they have now to be joined in a grasp plan. Area AIP is in charge of transforming the visual data provided by CIP and other areas into an appropriate hand configuration for grasping the target feature. The goal is to translate information about size, location and AOS/SOS representation into hand joint space. The specification of exact contact locations is not necessary, the finger placing and the movement trajectory being dependent on contextual optimization of the quality criteria. Grasp planning can be performed following a short sequence of logical steps.

The task-driven decision on the type of grip to perform (precision or power) is taken in advance. If the goal is to perform a power grip, visual analysis is usually very simplified (as suggested by the reduced activity in AIP), and only the object center of mass has to be approximately calculated. In this case, the hand has to move toward the object center, and the opposition axis of the fingers on the object is determined by the motor cost, which is minimized avoiding unnecessary rotations. The reaching action that requires minimum joint movements is thus executed, and once the palm gets in contact with the objects, the fingers close around it.

For precision grips, the requirements are different according to the AOS/SOS coding. If the object has a prevailing AOS activation, and the long axis is free for grasping, like for a standing cylinder, there is no preferential approaching direction apart from the one provided by the hand pose. The grasp action will be performed so that the opposition axis between thumb and fingers passes close to the center of mass, maximizing the correspondent criterion, and in a way that minimizes the cost of the movement, maintaining the trajectory as straight as possible and avoiding unnecessary rotations (Fig. 6.11a). If the object with prevailing AOS activation is laying on its long axis then wrist rotation is required, as only one approaching direction carries to the correct grasping position, from above and toward the center of mass (Fig. 6.11b). In the former case, an involving grip which includes contact between object and hand palm can be executed if required, in the latter case only fingertip grips are possible.

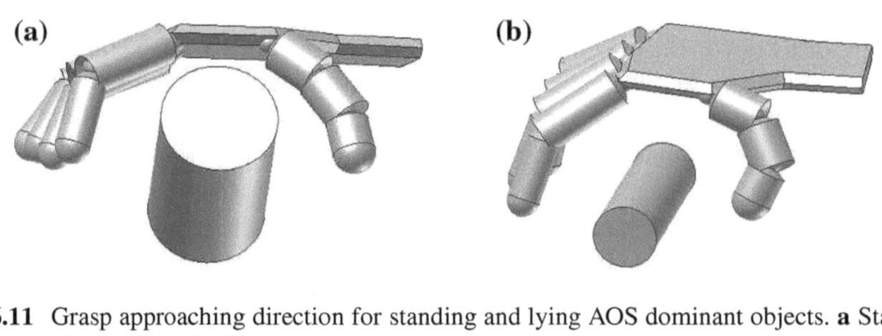

Fig. 6.11 Grasp approaching direction for standing and lying AOS dominant objects. **a** Standing AOS object. **b** Lying AOS object

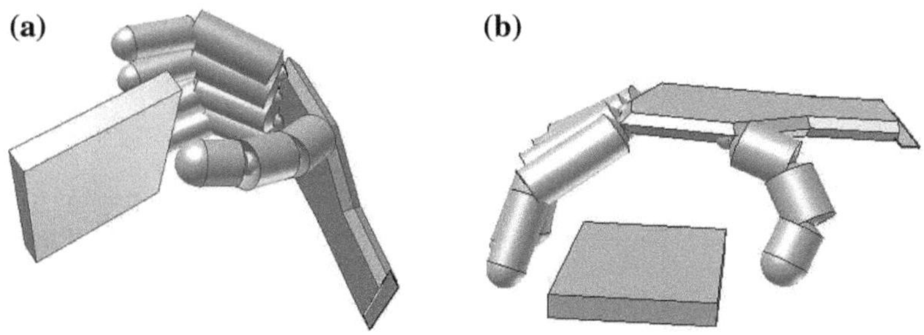

Fig. 6.12 Grasp approaching direction for standing and lying SOS dominant objects. **a** Standing SOS object. **b** Lying SOS object

If the object has a prevailing SOS activation, and the thin, grasping dimension is free, the direction of grasping is the one which makes the fingers oppose on the minor dimension of the object. The final part of the reaching movement is constrained to a plane, and there is still one degree of freedom for optimizing movement cost and center of mass approaching. Both visual and motor criteria have thus to be taken into account. If there are no other constraints, it is safer in this case to grasp an object from above and not from the side, in order to minimize the effect of gravitational torques. For light objects destabilizing torques are unlikely, and reaching comfort prevails (Fig. 6.12a). Object identification would help in this case. If the object is laying on its preferred graspable feature, the grasping action will have to be performed on a different dimension, along the main inertia axis (Fig. 6.12b), or not performed at all.

As can be observed in Fig. 6.9, some shapes show no clear dominance of either SOS or AOS activation. Objects with this characteristic can be grasped with both strategies, and again movement economy can be the determinant factor for establishing the final grasping position. Activation thresholds could be employed to distinguish the cases of SOS dominant, AOS dominant and neutral objects.

The described procedure leaves out the cases of objects that are approximately spherical, or that simply do not offer clearly graspable axes or surfaces. In such cases, a grasp in which fingers are abducted is preferable, to distribute the grasping force around the object surface. Most commonly in these situations, the ring and small fingers are used just for providing support and additional stability, whilst the index and medium fingers create a triangular force distribution with the thumb (see Fig. 6.13). For this reason the resulting grasp is called the *tripod grasp*.

Human grasping experimental results are consistent with a two virtual-fingers, hierarchical model of the control of a tripod grasp. First, an opposition space is selected between the thumb and the index and medium fingers, which form together the second virtual finger. Index-medium finger abduction depends on the object size (Gentilucci et al. 2003), and the direction of the forces exerted by the two fingers are symmetric with respect to the opposition axis (Baud-Bovy and Soechting 2001). If the object is irregular, tactile feedback is required in order to adjust finger position and force distribution in order to find a stable configuration before lifting the object. fMRI research support a substantial identity in the processing of two finger precision

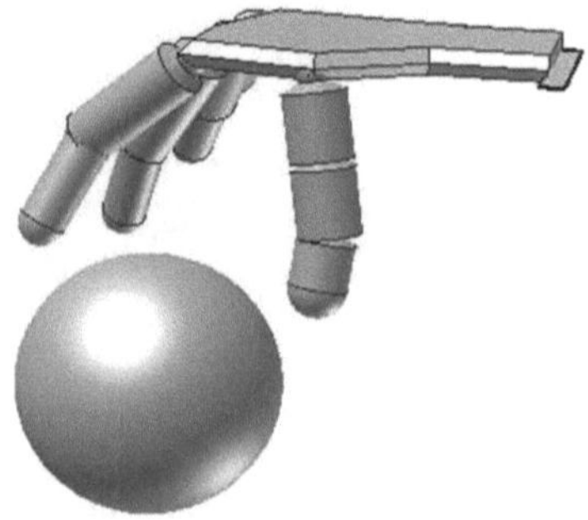

Fig. 6.13 Tripod grasping on a spherical object

grasps, tripod grasps and extended tripod grasps in which all five fingers contact the object (Cavina-Pratesi et al. 2007).

The above description, although valid for robotic implementation, is just a qualitative description of the results of AIP processing. Suggestions for an implementation closer to the cortical mechanisms are provided in Sect. 7.1.

6.2.4 Grasp Execution

The final grasping action is executed following the above guidelines, making use of visual information regarding object pose and location, and taking into account relevant grasp quality criteria. The grasping system introduced in Sect. 5.4.1 allows for the execution of differently shaped precision grips, including the tripod grasp. Power grips, although possible, are not controllable and thus avoided, as the hand palm is not endowed with tactile sensors necessary to detect the contact with the object. A wrong positioning could hence result in an excessive solicitation of the hand, with risk of damaging it.

6.2.4.1 The Reach and Grasp Movement

Before the movement onset, the goal position and direction of the opposition axis are defined as described in the previous section, and computed using the estimated location, pose and size of the object. The initial position of the arm, corresponding to the fixation period, before the movement starts, is shown in Fig. 6.14a. The first part of the reaching movement is just aimed at reducing the distance between object and effector. The final stage of the reaching action is more precise, and has to be performed moving perpendicularly to the opposition axis, from a short distance to the goal position. A *via posture*, i.e. an intermediate position and orientation goal (Meulenbroek et al. 2001), is defined in order to allow the correct execution of such stage, ensuring at the same time that no collisions with the target object are possible. The intermediate goal position has the hand in the correct grasping

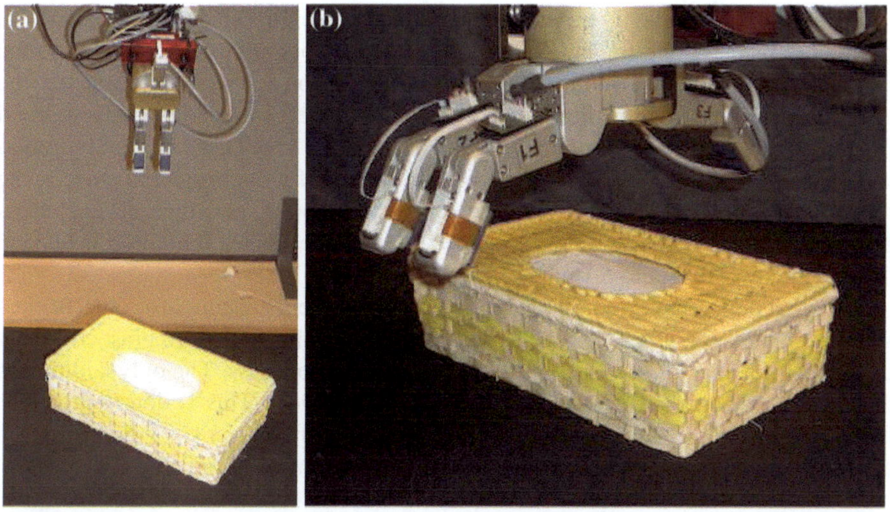

Fig. 6.14 First stages of grasping action execution, during object fixation before movement onset, and during reaching, before the hand closes on the object. **a** Fixation position. **b** Via-posture

direction, and the distance from the object is such that movements different from the approaching one would not result in unwanted contacts. Figure 6.14b shows an example of appropriate via-posture. A safety margin is also added to the expected object size to compensate for possible estimation errors. This is consistent with the findings of Hu et al. (1999), suggesting that hand preshape is performed taking into account also the object dimensions not directly involved in grasping. Such dimensions affect the security of the transport movement, which could be at risk of collision if during hand approaching the fingers pass too close to the object.

During the first stage of the transport movement the hand rotates toward the correct orientation while the arm moves to the intermediate position. Once the via-posture is reached, the hand is in the correct direction and, without stopping the movement, the robot arm reaches further toward the object until the fingertips are at level with the estimated object center of mass (Fig. 6.15a). Until this point the process

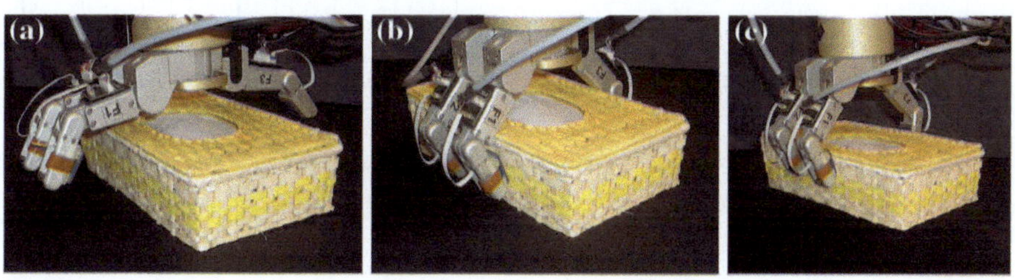

Fig. 6.15 Last stages of grasping action execution, during finger closing and object lifting. **a** Closing position. **b** Grasping. **c** Lifting

is fully open-loop, and only driven by the initial grasp plan. Once the estimated final position is reached, the fingers close, and tactile sensors are used to determine the moment of contact between fingertips and object. As soon as a contact is detected, the corresponding finger stops moving. The grasping movement is completed when all fingertips have contacted the object, as depicted in Fig. 6.15b. The grasp configuration is then checked, as described in the next section, and if it is considered correct, the object is lifted (Fig. 6.15c). If a finger misses the object, and thus no contact is ever detected, it stops as a security measure when it reaches a minimum extension threshold, currently set to 54 mm.

For what concerns tripod grasping, it is at the moment executed for objects classified as spheres (Sect. 5.4.2), which are approached from above, and grasped so that the opposition axis passes ideally through their center of mass. The only addition to the usual procedure is the separation between the fingers. This is done setting the finger opening angle θ (see Fig. 5.11) proportional to the object size: the bigger the object, the higher the finger separation. The opening angle, in radians, is given by:

$$\theta = \frac{D - i}{D} \tag{6.11}$$

where D is the object diameter and i the inter-finger distance (see Fig. 6.19). The outcome of using (6.11) can be observed in Fig. 6.16, in which finger positions and force directions for spheres of different sizes are shown. This solution allows to homogeneously distribute the contact points around the shape while maintaining the force directions as orthogonal as possible to the object side. In all other cases of normal opposition grips, in which fingers e_2 and e_3 are parallel, $\theta = 0$.

Examples of via-posture and grasp execution for a vertically placed cylindrical shape and a spherical shape are shown in Fig. 6.17.

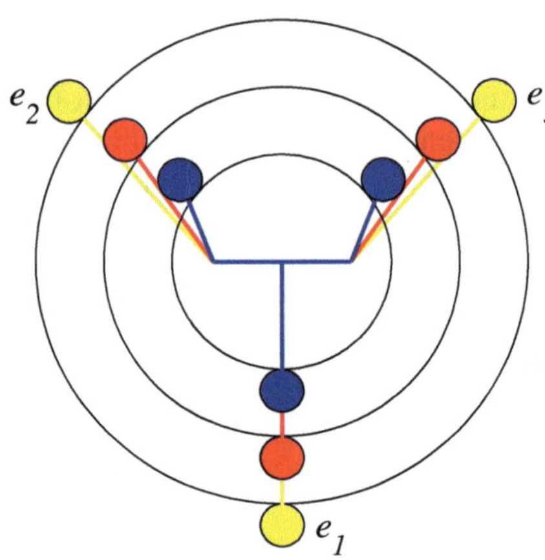

Fig. 6.16 Finger positions and force directions obtained with expression (6.11) for grasping spheres of different sizes (80, 130, 180 mm)

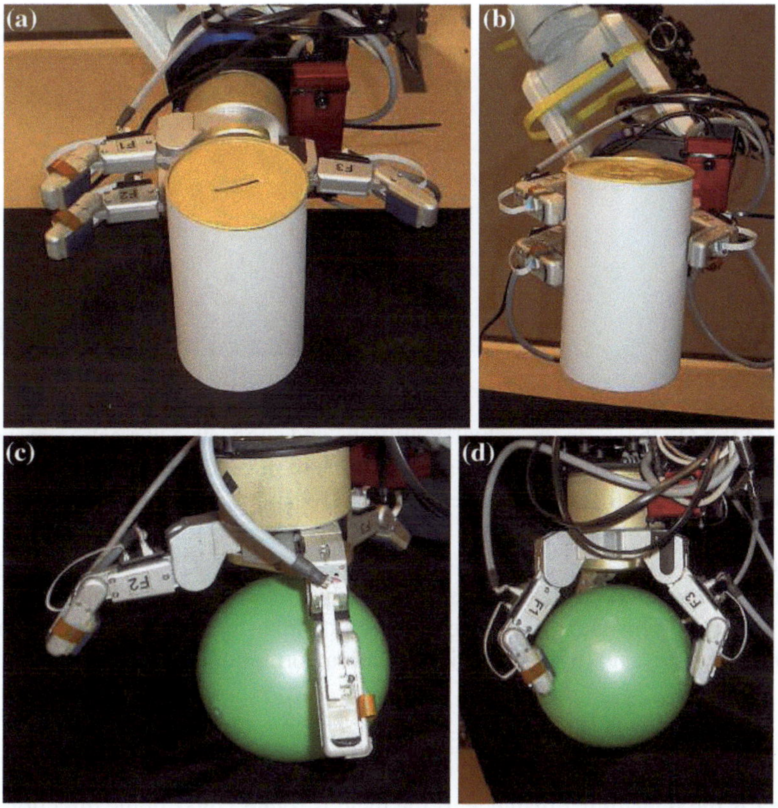

Fig. 6.17 Via-postures and grasp execution for a vertically placed cylindrical shape and a spherical shape. **a** Via-posture for cylinder. **b** Cylinder grasping. **c** Via-posture for sphere. **d** Sphere grasping

6.2.4.2 A Helping Tactile Hand

When the fingers close on the object, they stop at the moment of contacting the object surface. If the pose estimation was correct, the movement performed as required, and the object did not move, the fingers should present a substantial equilibrium in their final extensions. The expected proprioceptive state of the hand, corresponding to a symmetric grasp with respect to the object grasping axis, is of equal extensions for the three fingers. This constitutes a basic forward model of the expected action outcome. If, for a divergence between the estimated and the real values of distance, pose and size of the object, or for any other unexpected factor, the fingers touch the object with different extensions and orientations, the grasp could be unstable (see Fig. 6.18). In these cases, differences between finger extensions are detected, and proprioceptive hand feedback is used to adjust the grasping action to the real conditions, and thus achieve the necessary grip stability. An adaptation of the finger extension criterion

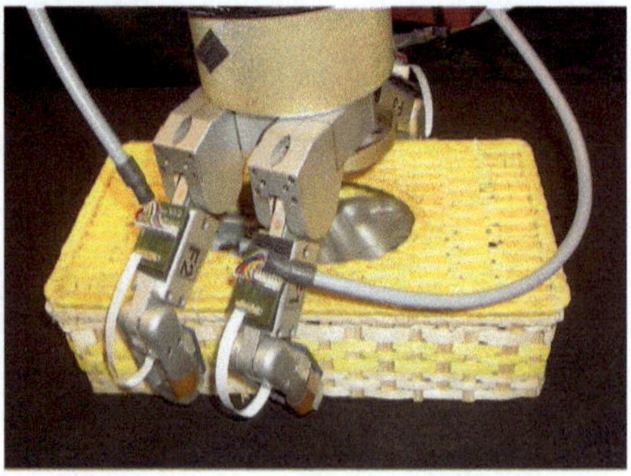

Fig. 6.18 Example of unstable grasp requiring a correction movement

provides the feedback on the actual conditions and suggests a correction movement if necessary. A proper action for adapting the hand pose to the new situation can be computed from the difference between finger extensions. As represented in Fig. 6.19, any correction movement is made of two components: z for translation and α for rotation.

Orientation correction is necessary when the orientation of the object is different from expected. This situation is identified by comparing the extensions of the two parallel fingers e_2 and e_3. The required rotation correction α is given by:

$$\alpha = arctan\frac{e_3 - e_2}{i} \tag{6.12}$$

where i is the inter-finger distance. If the hand is rotated by α in the direction of the finger with shorter extension, it will get in the situation in which the parallel fingers should contact the object with the same extension. A threshold is used to allow for minimum unavoidable extension differences which do not affect grasp stability. With the current settings, the correction movement is executed only if $|\alpha| > 2°$.

Translation correction is performed when the position of the object is different from the estimation. This case corresponds to a difference in extension between the thumb and the opposing fingers. The thumb extension is compared to the average extension of the opposing fingers, taking into account also possible extension differences between them. The displacement required for position correction is computed with the following expression (see again Fig. 6.19):

$$z = \frac{1}{2}(\frac{e_2 + e_3}{2} - e_1) \tag{6.13}$$

Again, this is the displacement that would carry the hand to the planned grasping axis. The threshold for moving is set at $|z| > 5$ mm. This threshold value and the one for rotation correction could be more appropriately set through a learning framework driven by the results of experiments performed in different conditions. Different thresholds could also be used for different objects.

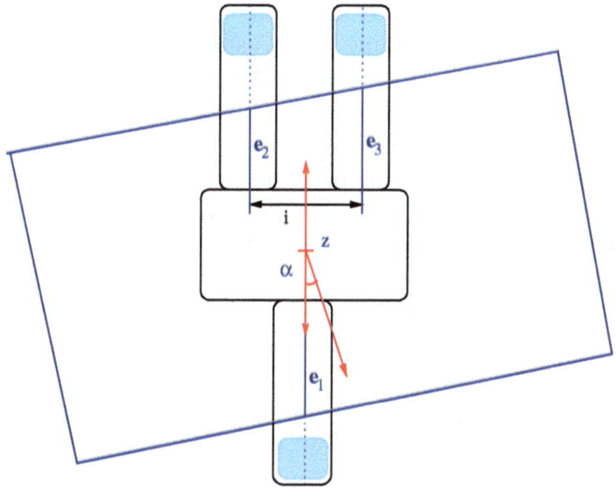

Fig. 6.19 Representation of the correction movements in rotation, α, and translation, z

During action execution, once all fingers have contacted the object and stopped moving, (6.12) and (6.13) are computed. If at least one of them is above threshold, the fingers open again, and the movement that compensates for the required orientation and translation correction is calculated and executed. If needed, the process repeats until no other corrections are required, then the fingers close firmly on the object and lift it.

The described grasping technique has been tested in two different conditions, i.e., without or with object displacement. The first condition, corresponding to a normal working situation, usually ends with a successful grasping action without performing any correction movement. In fact, in almost all cases the input provided by the visual system is good enough to allow the execution of the grasping action without the need of correcting hand position or orientation.

During the second type of tests, in perturbed conditions, changes in the object position and/or orientation were introduced on purpose, to check if the system was able to deal with unexpected and suddenly changing situations. The changes were applied after the visual analysis had been finished so that the real pose of the object was different from the estimated one, like in the example of Fig. 6.18. In this situations the robot might not be able to grasp the object without the support of the tactile feedback. Using information about finger extensions and hand contacts with the object surface, hand orientation and position are corrected as described above. When the difference between the real and estimated object pose is big, more than one correction movement might be required.

This framework presents two major limits. The first is that only displacement errors parallel to the grasping axis can be corrected. Any deviations from the object center of mass along the other two directions will not be detected, unless one of the fingers misses the contact. The second problem is with objects which edges are not parallel. In such cases, a rotation correction movement will be performed although not required, as α is always above threshold. A possible solution is to increase the threshold to a trade-off value which does not affect the reliability of normal grasping actions and, at the same time, is suitable for many unusual conditions.

This correction method models, in a simple way and at high level, the comparison between expected and real somatosensory input, described in Sect. 4.3.4, and determines a correction movement that aims at reducing such difference. In this way, grasp stability is implicitly achieved, through minimization of the difference between detected proprioceptive state and expected goal state given by a basic forward model. Following a simplified version of the schema of Demiris and Hayes (2002), very simple inverse models (Eqs. 6.12 and 6.13) compare the goal state of the hand with its actual condition and generate a motor command suitable to approach the goal. The forward model evaluates the outcome of the current motor command and guides the following step in order to keep improving the quality of the ongoing situation estimated by the finger extension criterion.

6.3 Conclusions

Compared to its neighbor grasping area AIP, the knowledge, and especially the modeling regarding one of the most fundamental areas of the dorsal stream, the posterior intraparietal area CIP, remains relatively undeveloped. In the first part of this chapter, a detailed analytical interpretation of CIP tasks is provided which takes into account both the computational and the neurophysiological points of view. The coding of visual features as it is thought to be performed by CIP neurons is employed in the second part of the chapter for generating appropriate grasp configurations. The integration of different kinds of grasp-related information and constraints, as performed by area AIP, is modeled and adapted to the requirements of the robotic system. Grasping experiments, performed with the aid of tactile feedback, confirm the suitability of the model to real robotic setups.

Neuroscientific plausibility and practical usefulness of the proposed vision-based grasping model have been justified, but several directions for further development in both regards can be devised. The next chapter will present a number of optional developments and required improvements for the robotic application. Issues regarding necessary refinements and possible alternatives for the model are also discussed.

References

Baud-Bovy G, Soechting JF (2001) Two virtual fingers in the control of the tripod grasp. J Neurophysiol 86(2):604–615

Bingham GP, Muchisky MM (1993b) Center of mass perception: perturbation of symmetry. Percep Psychophys 54(5):633–639

Borst C, Fischer M, Hirzinger G (2004) Grasp planning: how to choose a suitable task space. In: IEEE international conference on robotics and automation, New Orleans, USA, pp 319–325

Cavina-Pratesi C, Monaco S, McAdam T, Milner D, Schenk T, Culham JC (2007) Which aspects of hand-preshaping does human AIP compute during visually guided actions? Evidence from event-related fMRI. In: Annual meeting of the Society for Neuroscience

Chinellato E, Morales A, Fisher RB, del Pobil AP (2005) Visual quality measures for characterizing planar robot grasps. IEEE Trans Syst Man Cybern Part C 35(1):30–41

Chinellato E, Morales A, Sanz PJ, del Pobil AP (2003) Validation of features for characterizing robot grasps. In: Mira J, Alvarez J (eds) Artificial neural nets problem solving methods, LNCS 2687, Springer, pp 193–200

Clower DM, Dum RP, Strick PL (2005) Basal ganglia and cerebellar inputs to 'AIP'. Cereb Cortex 15(7):913–920. doi:10.1093/cercor/bhh190

Demiris Y, Hayes G (2002) Imitation as a dual-route process featuring predictive and learning components: a biologically-plausible computational model. In: Dautenhahn K, Nehaniv C (eds) Imitation in animals and artifacts, Chap 13. MIT Press, Cambridge, pp 327–361

Doya K (1999) What are the computations of the cerebellum, the basal ganglia and the cerebral cortex? Neural Netw 12(7–8):961–974

Eastough D, Edwards MG (2007) Movement kinematics in prehension are affected by grasping objects of different mass. Exp Brain Res 176(1):193–198. doi:10.1007/s00221-006-0749-3

Fukuda H, Fukumura N, Katayama M, Uno Y (2000) Relation between object recognition and formation of hand shape: a computational approach to human grasping movements. Syst Comput Jpn 31(12):11–22

Gentilucci M, Caselli L, Secchi C (2003) Finger control in the tripod grasp. Exp Brain Res 149(3):351–360. doi:10.1007/s00221-002-1359-3

Hu Y, Eagleson R, Goodale MA (1999) The effects of delay on the kinematics of grasping. Exp Brain Res 126(1):109–116

Hu Y, Osu R, Okada M, Goodale MA, Kawato M (2005) A model of the coupling between grip aperture and hand transport during human prehension. Exp Brain Res 167(2):301–304. doi:10.1007/s00221-005-0111-1

Jenmalm P, Dahlstedt S, Johansson RS (2000) Visual and tactile information about object-curvature control fingertip forces and grasp kinematics in human dexterous manipulation. J Neurophysiol 84(6):2984–2997

Meulenbroek RG, Rosenbaum DA, Jansen C, Vaughan J, Vogt S (2001) Multijoint grasping movements. Simulated and observed effects of object location, object size, and initial aperture. Exp Brain Res 138(2):219–234

Morales A, Chinellato E, Fagg AH, del Pobil AP (2004) Using experience for assessing grasp reliability. Int J Humanoid Rob 1(4):671–691

Naganuma T, Nose I, Inoue K, Takemoto A, Katsuyama N, Taira M (2005) Information processing of geometrical features of a surface based on binocular disparity cues: an fMRI study. Neurosci Res 51(2):147–155. doi:10.1016/j.neures.2004.10.009

Roa M, Surez R (2015) Grasp quality measures: review and performance. Auton Robots 38(1):65–88. doi:10.1007/s10514-014-9402-3

Sakata H, Taira M, Kusunoki M, Murata A, Tanaka Y (1997) The TINS lecture. The parietal association cortex in depth perception and visual control of hand action. Trends Neurosci 20(8):350–357

Sakata H, Taira M, Kusunoki M, Murata A, Tanaka Y, Tsutsui K (1998) Neural coding of 3D features of objects for hand action in the parietal cortex of the monkey. Philos Trans R Soc B: Biol Sci 353(1373):1363–1373

Sakata H, Tsutsui KI, Taira M (2005) Toward an understanding of the neural processing for 3D shape perception. Neuropsychologia 43(2):151–161. doi:10.1016/j.neuropsychologia.2004.11.003

Shikata E, Tanaka Y, Nakamura H, Taira M, Sakata H (1996) Selectivity of the parietal visual neurones in 3D orientation of surface of stereoscopic stimuli. Neuroreport 7(14):2389–2394

Chapter 7
An Ever-Developing Research Framework

This chapter presents a number of issues and developments that could not fit in the presentation of the model in Chaps. 4, 5 and 6. Some aspects are controversial and need additional neuroscience data to be modeled, others would need a different robotic setup to be properly tested. Some ideas have simply not been implemented yet, or have been implemented only partially, and thus do not fit in the main modeling framework.

In Sect. 7.1, the sort of visuomotor transformations performed by AIP are further discussed and suggestions for additional modeling provided. Section 7.2 presents a detailed modeling proposal for the interaction between the streams. A research work performed on the subject in collaboration with Blaise Pascal University is summarized. Finally, in Sect. 7.3, the fRI interface for visualizing the activity of a robot as in a simulated fMRI experiment is presented. In general, implications for further studies in neuroscience, computational modeling and robotic applications are discussed.

7.1 Purely Visual and Visuomotor Transformations in AIP

The proposal for generating suitable hand configurations from visual data advanced in Sect. 6.2 is constrained by the implementation of other modules and by limitations of the robotic hardware. The job of AIP in grasp planning and execution is surely more complex and nuanced. All ventral stream information, such as weight or friction, and task related issues are left aside for the moment, and discussed in Sect. 7.2.

7.1.1 Visual-Visual Transformations

Object type neurons in AIP seem to encode 3D object geometry in a grasp-oriented way (see Sect. 2.3.2). Following the suggestions of Rizzolatti and Luppino (2001)

E. Chinellato and A.P. del Pobil, *The Visual Neuroscience of Robotic Grasping*,
Cognitive Systems Monographs 28, DOI 10.1007/978-3-319-20303-4_7

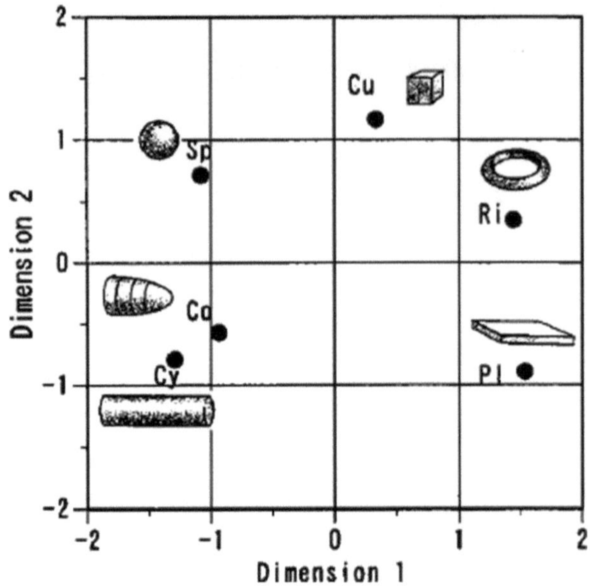

Fig. 7.1 Principal components of visual object representation for object type neurons of the anterior intraparietal sulcus AIP. From Murata et al. (2000)

and the experiments of Fukuda et al. (2000), they should perform a purely visual transformation of the SOS/AOS coding received from CIP. Figure 7.1, taken from Murata et al. (2000), represents the two principal components of the activation of AIP object type neurons during fixation of different target objects.

The authors performed the same mapping for object type neurons during grasping, and for non-object and motor type neurons. The first is very similar to the one displayed in Fig. 7.1, the other two are different but very close to each other. The shape distribution may be interpreted in many ways, but a first qualitative analysis suggests that the AOS/SOS distinction is approximately represented in the first dimension. It is less clear what object qualities are discerned by the second axis. There seems to be an important symmetry component, but also a curvature effect. Moreover, it is not clear how the extremely important aspect of object size enters the schema. In any case, there is in principle a direct mapping between this representation and the input coming from CIP.

A second, more complex transformation has to be performed for matching visual properties to hand configurations.

7.1.2 Visuomotor Transformations

The above data refer to experiments with macaques. Although the basic concepts are not expected to change, the greater dexterity of the human hand and the increased complexity of the human intraparietal sulcus suggest that the situation is more com-

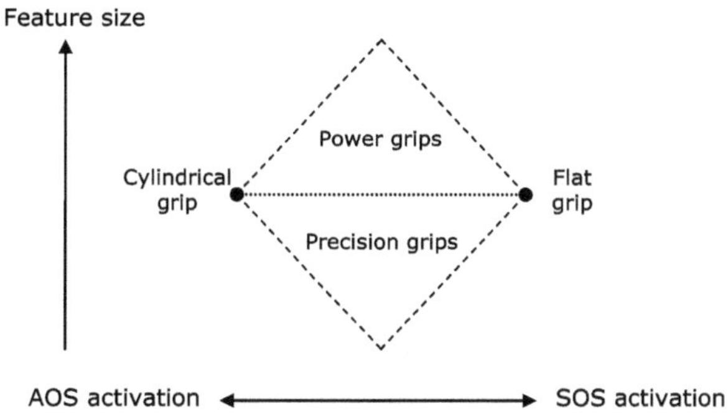

Fig. 7.2 A basic 2D space of grip taxonomy

plex for human grasping. It is reasonable to assume that a representation similar to the SOS/AOS, although somewhat more elaborated, is used in the human IPS as well. Similarly, the graphs of Murata et al. (2000) probably show a simplified version of what could be found in human AIP. A simple proposal on which to build a theory of visuomotor transformations in human AIP is provided below.

In Fig. 7.2, a 2D space is depicted which jointly represents SOS and AOS activations and a basic grasp taxonomy. Such taxonomy is very likely a multidimensional joint space described by all possible combinations of joint angles, in which only the extreme cases can be clearly identified with recognizable basic hand shapes. In the proposal of Fig. 7.2 the basic grips of the taxonomy are, on the vertical direction, the full-hand or power grip and the smallest pinch or precision grip. In the horizontal direction, there are the ideal flat grip in which only the metacarpal-finger phalanges are flexed, so that all fingers are parallel and opposed to the thumb, and the ideal involving or cylindrical grip, in which all phalanges are flexed and the hand closed in a fist. These two last grips correspond to the ideal maximum activation of SOS and AOS neurons, as the former is suitable to grasp a perfectly flat surface, and the latter a thin elongated shape.

Summarizing, the list of theoretical *extreme* grips from which all hand configurations derive are:

flat grip: SOS neurons are dominant; predominant flexion of proximal phalanges, pad opposition between thumb and other fingers;

cylindrical grip: AOS neurons are dominant; all phalanges are flexed, involving the object, finger pad and palm opposition;

precision grip: strong activation of one or both kinds of orientation selective neurons; adapt for small features requiring thorough visual analysis, the thumb is opposing the index or middle finger, the other fingers can be used for support;

power grip: limited activation of SOS and AOS neurons; full hand, spherical grip, maximum contact of fingers and palm, used for default when no other grips are considered appropriate, or if visual analysis is considered unreliable or incomplete.

All possible grasping configurations would thus be found inside the parallelogram, classified either as a precision or a power grip, with a continuous separation between the two, and also somewhere in between the involving and flat grips. Indeed, the four ideal grasps are just abstractions, and it is reasonable to assume that the edges and corners of the parallelogram do not correspond to real grasping configurations. Moreover, whilst "pure" cylindrical grips are more easily associated to power tasks, opposition grips appear more dexterous, suggesting that there is no real symmetry as the graph would indicate.

From an ecological point of view, the plausibility of the proposed representation is supported by a sort of economy principle for grasping, according to which more general purpose and less precise grips do not need detailed visual analysis. In fact, small, precision grips probably need detailed information from both kinds of neurons, whilst power, full hand grips are likely to represent a default choice.

Unfortunately, this proposal cannot be tested at the current stage of development of the model. Firstly, as already pointed out, curvature and size coding of CIP neurons have not been properly characterized yet. Then, AIP studies did not analyze the same objects used for CIP. It would be very helpful for example if cylinders and parallelepipeds of the same size and proportion could be presented to CIP and AIP and their activations compared, to study the effect of curvature changes on the two areas. The same should be done for other types of objects, and the effect of size should be studied in a similar way.

Further limitations are posed by the robotic hand. In fact, the Barrett Hand does not allow for voluntary, independent movements of the finger phalanges, so that, unless for very large objects, the grip is practically always cylindrical.Moreover, palm opposition can not be performed properly, because there are no touch sensors on the palm which would detect object contact.

7.1.3 The Reaching and Grasping Action

The best solution to construct the visuomotor matching between visual data and hand shapes is probably to develop a robotic experimental framework in which suitability of hand shapes to visual data is gradually learnt, for example through a reinforcement learning process in which grasp stability is positively rewarded. Such framework would allow to analyze naturally emerging visuomotor transformations, and test the efficiency of different alternative visual representations. Bootstrapping for the learning process could be provided by basic postures identified through examples of human joint spaces extracted with a data glove.

The same postures could constitute an ideal starting point for building truly human-based motor primitives for the robot arm and hand. Lim et al. (2005) extract movement primitives from the analysis of human movements, and propose a method for joining them and implementing motions in robotic applications. Their guidelines, together with the experience provided by other related works (see e.g. Nori and Frezza 2005), could constitute the starting point for merging accurate visual analysis with biologically plausible action execution. A framework of this kind would also be particularly suitable for monitoring of action execution through an implementation of the forward model/inverse model principle more accurate than the one proposed in Sect. 6.2.4.2. Such implementation could be done following the guidelines of Kawato (1999), Wolpert and Ghahramani (2000) and Miall (2003) and forward models can be learnt from experience as in Dearden and Demiris (2005).

Timing and coordination between action components, which are not dealt with in the model, would benefit from this sort of modeling improvement. For example, the correct coupling between the reaching and grasping movements is an issue that has not been considered, as often happens in robotics applications. This is instead a fundamental and largely studied aspect in human grasping, and various plausible models on the relation between reaching and preshaping have been developed (Shadmehr and Wise 2005). The hypothesis of parallel visuomotor channels for the transport and the preshaping components of the reach-to-grasp action is well recognized (Jeannerod 1999). Alternative models, such as the *multiple finger reaching* idea (Smeets et al. 2002), are not given much credit, due to the quantity and quality of evidence supporting the mainstream hypothesis (see e.g. Tanné-Gariépy et al. 2002; van de Kamp and Zaal 2007). In any case, the coupling between the two subsystems linking parietal and premotor cortex is tight, and they must share a common mechanism for coordinating with each other (Jeannerod 1999; Roy et al. 2002; Fattori et al. 2010).Various computational models of reaching and grasping coordination that might be suited to robotic implementation are available (Mon-Williams and Tresilian 2001; Jiang et al. 2002; Ulloa and Bullock 2003; Hu et al. 2005). Primates often perform coordinated eye and arm movements, contextually fixating and reaching towards nearby objects. Taking inspiration from such visuomotor mechanisms (Chinellato et al. 2010; Bosco et al. 2009), (Chinellato et al. 2011) designed and developed a model for learning and producing the coordinated control of gazing and reaching movements.

7.1.4 After Contact

A more immediate extension to the developed framework is the use of post-contact information for improving the reliability of the vision modules. In fact, after tactile adjustment, the exact position, orientation, and one of the object dimensions can be exactly measured, through proprioceptive feedback on the hand and arm state. These values can be compared to the initial estimations, and the error magnitude and sign of each measure memorized, in association with the object class. Even more, errors can be calculated for each one of the different estimators presented in Sect. 5.4.3. In this way, for each estimator a reliability function dependent on object size, distance and pose can be defined, and used in the following experiments. Human experiments support this proposal, as it has been demonstrated that cue weighting in slant perception can be modified by haptic experience (Ernst et al. 2000). The next-generation estimator will thus perform cue integration using both correlation and reliability, as it is done in the primate brain.

Summarizing, an improved modeling of the visual and visuomotor transformations performed in AIP should consider learning processes in which multimodal tactile and visual feedback is used to bias the matching between different representation levels.

7.2 A Tighter Interaction Between the Streams

Even though it is widely accepted that the dorsal and ventral streams tightly interact with each other, the mechanisms underlying such interaction are still largely under debate (Goodale 2008; Cloutman 2013). In the neuroscience review, two levels of direct integration between the streams were mentioned, one for the ventral connections of CIP and the second for those of AIP.

7.2.1 Links Between CIP and the Ventral Stream

Section 2.3.1.3 described the possible bidirectional links of CIP with ventral stream areas (Zanon et al. 2010; Perry et al. 2014). The dorsal use of information regarding object class was explicitly modeled in Chap. 5. On the other hand, the second step in the framework of Sect. 5.2.3, full object recognition, is not dealt with in the implemented system, and can be simplified and accelerated by dorsal input. Object recognition could rely for example on a chain code invariant representation as the one initially used for classification, and which revealed more suitable for single object identification than for class discrimination (see Sect. 5.4.2). Various researchers pointed out though that action-related information maintained in the dorsal stream is likely to play an important role in the object recognition process, as a set of possible affordances constitutes an additional way of identifying an object (Sugio et al. 1999; Shmuelof and Zohary 2005). The SOS and AOS responsiveness found for the target object could be one possible format used by the dorsal stream to help the ventral areas in the recognition task. It is in fact very unlikely that two objects share the same SOS and AOS activations. CIP projections would thus provide the ventral stream with additional information for improving the reliability and speed of object recognition. For what concerns the representation of known objects, in their first years of development, human beings accumulate experience on properties such as color, texture, material, object identity, learning the likelihood of different relations among them. A working model of this recognition and generalization capacity should rely on a knowledge base founded on these properties (see e.g. the proposal of Metzinger and Gallese 2003).

There is a second ventral → dorsal link that could involve intraparietal areas other than AIP. This projection would occur after object recognition, and would provide the dorsal stream with the exact object size as memorized in ventral areas. In fact, object identification would allow to recover specific stored knowledge on object dimensions that can be used to resolve more easily and accurately the size/distance ambiguity. This mechanism is supported by several different studies comparing grasping accuracy using monocular and binocular data and different kinds of objects (Loftus et al. 2004). For example, it has been shown that proprioceptive cues such as vergence and accommodation are relied upon especially for novel objects, suggesting that for known objects retinal data, inherently ambiguous otherwise, are predominant (Mon-Williams and Tresilian 1999).

7.2.2 Links Between AIP and the Ventral Stream

The second stage of integration between the streams refers to the direct links between ventral areas and AIP (Zanon et al. 2010; Verhoef et al. 2011), which have been repeatedly pointed out in the book. An explanation of how properties related to the object identity would affect the selection of grasp features and contact points have been provided in Sect. 6.2.2, but the proposal has not been implemented in all its details yet. As explained, a higher confidence in the object recognition/classification process reflects in a stronger influence of past grasping experiences, whilst a more uncertain recognition leads to a more exploratory behavior, giving more importance to actual observation. Recognition of object identity can affect dorsal processing at three levels.

The first level, described above, is the ventral contribution to object size and distance estimation. At the second level, the ventral stream can provide the dorsal with the exact object weight, roughness, compliance, and consequently orient the criteria employed for action selection, as described in Sect. 6.2.2. Hand preshape, target contact points and initial grasping force would be contextually defined according to the ventral information (Perry et al. 2014). The third and final level refers to the third stage of object recognition (Sect. 5.2.3), in which grasping has to be executed on a familiar object. It can be supposed that in this case the motor coding of a grasping action is already associated with the object. In fact, such patterns has been observed for tools, which seem to elicit dedicated learnt motor representations (Creem-Regehr and Lee 2005; Johnson-Frey et al. 2005; Valyear et al. 2007). While in the previous stages there was a variable balance between recalling and exploration, for this case the process of grasp planning is reduced to its minimum. Only object pose and distance have to be estimated, as the preshape movement, and the expected contact with the object are recovered from previous experiences.

The implementation of this mechanism has to be based on two modeling elements already cited above: an appropriate knowledge base of objects and a suitable vocabulary of motor primitives that can be associated to such objects. At this level of detail in the modeling of grasping behaviors, the task could not be disregarded, and the mentioned object knowledge base should include their possible utility. Therefore, different tasks can be associated with different motor programs, and a motor program can be associated to a given task even for different objects. If a hammer is not available, another tool will be used "as a hammer", emphasizing that the motor program is associated to the task, and can be transferred to novel objects. The final goal is a three-way model in which visuomotor coding, object identity and task requirements modulate the level of interaction between the streams (Plewan et al. 2012; Tankus and Fried 2012; Wokke et al. 2014).

The additional information provided by the tactile feedback upon contact of the hand with the object is critical in this process. The only way to learn the exact object size is through proprioceptive feedback after grasping, and the same happens for its weight and compliance. The above mentioned method for learning the reliability of the various visual cues in different conditions can hence be extended to associate with each object properties that are relevant for graping purposes. This can be achieved through a cross-modal object representation in which visual and tactile information complement each other, and which is directly related to the task to perform. A representation of this kind is likely maintained in the human LOC (Amedi et al. 2001; Binkofski et al. 2004).

7.3 fRI, Functional Robotic Imaging: Visualizing a Robot Brain

In this section, an auxiliary tool aimed at improving the integration between neuroscience, computational modeling and robotic systems is presented. This tool is the fRI (functional Robotic Imaging) interface for the functional visualization of a robot activity, described in detail in Chinellato and Pobil (2008).

The fRI interface associates each task performed by the robot to a brain area and visualizes its activation on a brain image, similarly to what happens in fMRI experiments (see text box on fMRI). The information flow can thus be monitored at a very high level, and especially non-roboticists can benefit from this kind of functional visualization. The scanning of the "robot brain" is much easier, faster and cheaper than real imaging, and far from representing an alternative to real experiments, it may help researchers in two important ways. First, fRI can be used to check in advance the appropriateness of experimental protocols, reducing expensive and complicate preliminary tests with human subjects (Culham 2006). Second, it represents a way for comparing, and validating, possible explanations of experimental results derived from different models. The ambitious goal is to make the tool useful for both the definition of experimental protocols and the interpretation of results according to alternative models. The tool is suitable for models that are applicable to robotic setups while maintaining a strong functional resemblance with brain areas. In this way, the rigorous model design is likely to represent an ideal inspiration source for planning completely new experiments.

An additional feature of the fRI tool and the associated modeling architecture is the possibility of performing "impossible experiments". As mentioned in Chap. 2, neuroscience theories often derive from observations done on neurally-impaired people. Logically, such experiments cannot be reproduced, neither it is possible to decide beforehand the type of brain damage on which to investigate. If properly implemented, a layout as the proposed one would allow for such critical experiments.

The fRI approach presents two main differences with the only related work in the literature (Arbib et al. 2000), which deals with large computational models and the possibility of simulating brain imaging experiments with them. First, the proposed modeling paradigm is focused on the function of computational modules, and not on their structure or on particular implementation techniques. The second fundamental difference is the application-oriented stance of the current approach, which is designed to be implemented on a robotic setup, and not only computationally.

The part of the model dedicated to extract and map visual features to hand configurations, exposed in Sect. 6.2 and further discussed in Sect. 7.1, is used as a first testbed for showing how the fRI instrument works, and how it may constitute an additional tool for depicting the functioning of a relatively complex neural model.

7.3.1 Modeling Requirements

There are a number of restrictions for a model in order to be coupled with the fRI visualization tool. First of all, it has to be a functional model, implemented at the proper level of abstraction (see Sect. 4.2), and composed of various modules. The modules should be provided with the flexibility necessary for simulating different behaviors and skills through re-configuration, and not through re-programming. Only in this way can they be used to test theories and assumptions. This goal can be achieved using a distributed architecture of interconnected modules, which have to be simple and robust in order to allow easy modification of the data flow.

Recruitment between modules is driven by a set of connectivity rules and by the current data flow. A module corresponds to a neuronal population performing a certain function, and is associated with a modeled brain area, which can be composed of one or more modules. For example, there may be two modules for SOS and AOS activations upon visual presentation of a target object, and both modules would belong to area CIP. The rules that determine the connectivity of the set of modules represent a model of how the brain performs a given task. The projection of SOS and AOS activations to an AIP module representing object type neurons is an example of connectivity between modules. The pattern of activation of the different modules will thus depend on the programmed connectivity and on the input set.

For the recruitment of a module, it may be necessary to have more modules calling on it at the same time. Also, some modules could facilitate the activity of a specific module, others could inhibit it. For example, in a grasping task, the module which computes the distance of the object would inhibit grasp planning if the object is out of reach.

7.3.2 The fRI Interface

The purpose of the fRI interface is to visualize the actions performed by a robot as they were executed by a human subject during an fMRI experiment. The activity of the robot is thus displayed through a series of activations of brain areas, in a simulated fMRI analysis screen. The result is similar to what can be obtained after data analysis in a real fMRI experiment, like the one shown in Fig. 7.3.

From the list of implemented modules, and the relative brain areas, a library of locations is created, representing the position in the brain of the areas corresponding to the various modules available. Locations can be obtained from the literature, and are kept in Talairach coordinates, which allow an easy mapping on normalized anatomical images (Talairach and Tournoux 1988). For example, the anterior intraparietal sulcus AIP has been located with a 95 % confidence interval by averaging the results of different studies (Frey et al. 2005). Figure 7.3 represents the contrast

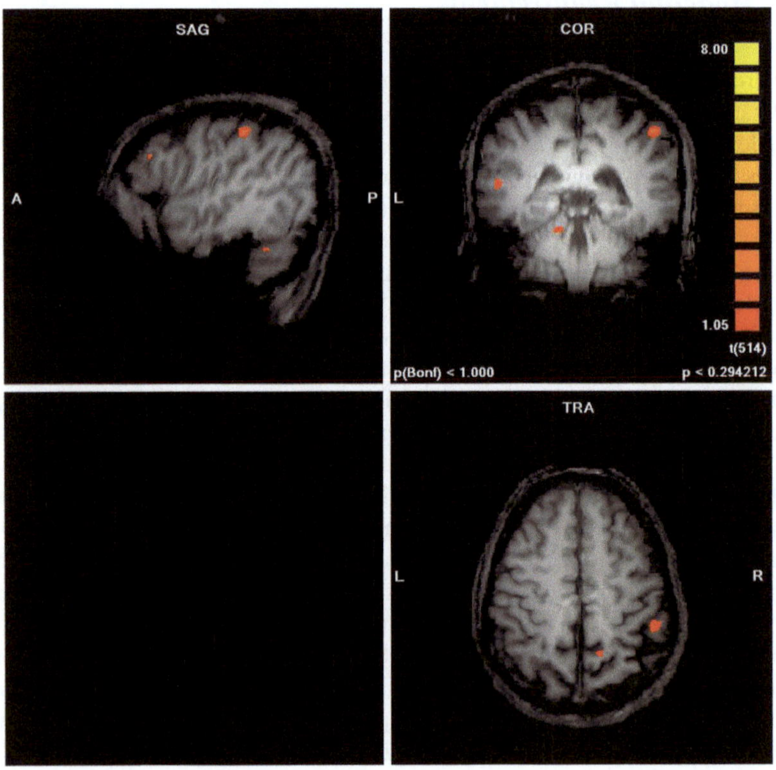

Fig. 7.3 Example of real fMRI activation, the larger activated area is the right anterior intraparietal sulcus (AIP). The views are as follows. *Top-left image* sagittal view (from the *left side* of the head); *top-right image* coronal view (from the back of the head); *bottom-right image* transversal view (from the *top* of the head)

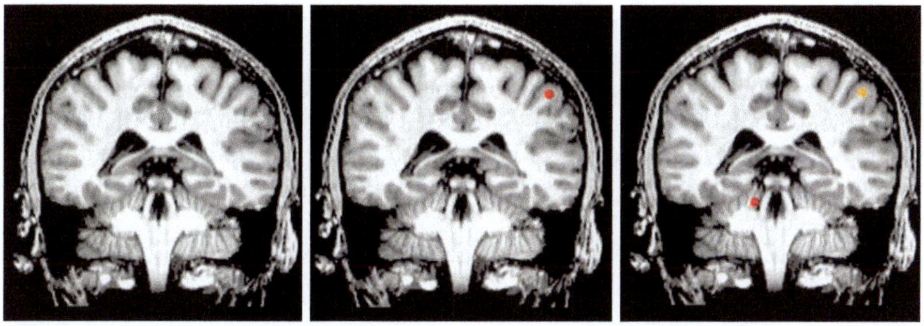

Fig. 7.4 Simulated fRI activation; no activation on the *left*, right AIP activation in the center, right AIP and left cerebellum activations on the *right*

obtained in an fMRI grasping experiment, in which the strongest of the visible activations corresponds to the right AIP. Using a high-resolution anatomical MRI scan as base, and extracting from it the required layers, activations can be superimposed and visualized by color-coding the coordinates corresponding to the areas that are active during the execution of a certain task. Being an area composed of one or more modules, its level of activation depends on the number of modules which are active at a given moment.

In Fig. 7.4 an example of the simulated functional imaging is given: a coronal layer of an anatomical scan is shown on the left with no activations, in the center with a superposed activation in the right AIP area, and on the right with two activations, one again in AIP, but with a lighter color and thus stronger, and another one in the left side of the cerebellum.

The dynamical behavior of the modules during a given task should be such that the pattern of activations is not known beforehand. Only in this way, the visualization of the activity of the "robot brain" allows to make a prediction of how a human model of connectivity would behave in a certain case (i.e. with a given input set). The choices taken during the modeling process will thus reflect in the activation observed in the simulated fMRI environment.

7.3.3 Reproducing and Predicting Experiments

The first step for fine-tuning the connectivity methods and test the real value of the tool is to use it for replicating real fMRI experiments. This process starts from a set of modules which supposedly allows for the execution of a number of tasks. An fMRI experiment regarding one such tasks is then chosen from the literature. Different input sets, corresponding to the conditions of the experimental protocol, are defined, and the robot performs the given task with all possible inputs. The behavior of the robot during the experimental session as visualized in the fRI is thus expected to reflect the activation of the same areas of the brain as in the original fMRI experiment. Of course, only the subset of areas included in the model can be considered: it is not plausible to include all areas normally activated by a given experiment, as this would imply the use of a largely implausible model of the whole brain. The following step would be to predict the outcome of new experiments, using the fRI in conjunction with the appropriate model. An example of how this can be achieved is provided in the next section.

7.3.4 fRI of the Posterior Parietal Cortex

The literature of neuroscience studies on vision-based grasping offers various experiments which can be reproduced with the fRI platform. A very interesting case is the analysis of the different patterns of activation obtained during power and precision grips. So far, no extensive results were available until recent research showed that AIP is much more active during precision grips than for power grips (Begliomini et al. 2007; Cavina-Pratesi et al. 2007). Previously, only Ehrsson et al. (2000) performed this experiment in the dark, observing that precision grips activate AIP more than power grips. Results for CIP are still not available.

The starting point for the test is the mapping of CIP neuronal activation to the fundamental grip taxonomy described in Sect. 7.1. According to the described framework, precision grips should activate either one or both of SOS and AOS neurons, depending on the exact size and shape of the target feature, whilst full-hand power grips would not consistently activate any orientation selective neurons. This would reflect in a higher activation of CIP for precision grips compared to power grips. It is also very plausible that grasp planning in AIP is going to be much more complex for precision than for power grips, as in the latter case no elaborate analysis for finger positioning is necessary, which should reflect in a lower activation. Robot grasping experiments based on the model are thus very likely to show increasing activation in both CIP and AIP for precision compared to power grips. It would be interesting though to observe the modulation in the observed activation according to gradually changing shapes and sizes of the objects. Performing these same grasping experiments on a real fMRI setup with human subjects could provide a comparison for testing the appropriateness of the model.

For illustration purposes, Fig. 7.5 shows the activation pattern expected in this example: left AIP is shown on the left, left CIP on the right. The Tailarach coordinates of the areas are those provided by Shikata et al. (2003).

As anticipated above, this pattern of activity has recently been confirmed for AIP, but not yet for CIP, which is anyway a more controversial area in human studies. Future studies could thus assess the problem of more clearly identifying where in the human brain feature extraction is performed. Also, differential activations of AIP in various conditions can provide further cues on how exactly visual information is used for generating grasp configurations according to the model framework.

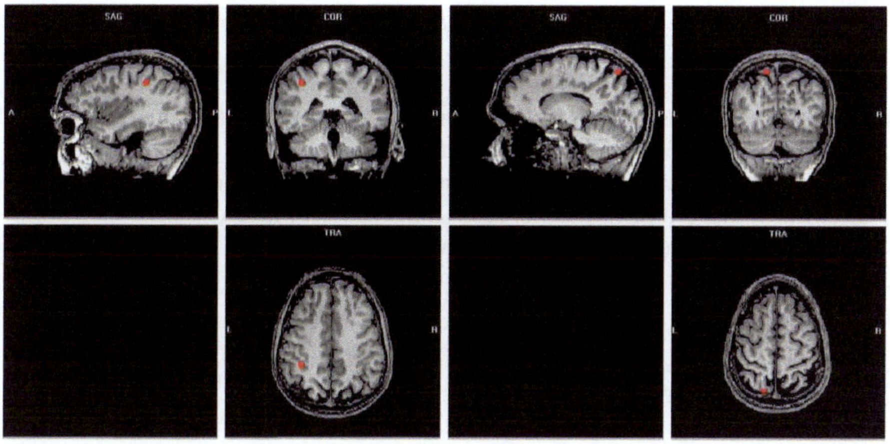

Fig. 7.5 fRI activations of left AIP (*left*) and left CIP (*right*)

7.3.5 The Brain-Damaged Robot

Brain lesions have probably been the best source of information for neuroscientists before the advent of brain imaging techniques. As mentioned in Chap. 2 the two streams theory, was first proposed by Goodale and Milner after observing the singular behavior of *visual agnosia* and *optic ataxia* patients (Goodale and Milner 1992). Reproducing the behavior of such patients can be an interesting test for the model and the fRI interface. This could be done in two complementary ways: modifying the connectivity of models, and introducing different quantities of noise in the data flow. The outcome may be used to determine which patterns of connectivity, when appropriately damaged, would actually generate the behavior observed in impaired humans. Nowadays, only transcranial magnetic stimulation (see text box on TMS) experiments can partially reproduce the effect of local brain lesions, but the tool is controversial as potentially unsafe, and it suffers from low resolution and limited access to a large part of the brain.

Several interesting experiments on brain lesion modeling could be performed. For example, optic ataxia, which is a consequence of dorsal stream damages, should strongly affect the grip generation and execution process, so that grasping actions may fail or being clumsily executed. Different effects would be produced by different connectivity changes, and hypothesis on exactly what parts of the dorsal stream may be affected can be made.

Computational modeling of brain lesions is not a completely new approach (Pouget and Sejnowski 2001; Salinas and Sejnowski 2001), and a model of impaired brain functions in grasping could be applied to a simulated environment. Nevertheless, the uncertainty of the real world can never be properly reproduced with only computational models, and even a simplified embodiment provides a different significance to the experimental validation. Implementation on a robotic setup surely requires an additional effort, but this is balanced by the more general value of the obtained results.

Summarizing, the proposed modeling approach and the fRI tool for interfacing computational models with robotic experiments aim at facilitating the interplay and the mutual positive influence between the research fields of robotics and neuroscience. The fRI interface can be used as an aid for the design of fMRI experiments and for the validation of functional brain models. Also, it can be used for "impossible experiments", as local temporary brain damaging, which cannot be performed on human subjects.

7.4 Further Extensions

In the previous sections, fundamental developments to the work presented in the book have been introduced and thoroughly discussed. There are other relevant aspects that have not been included in the current research. A few of them are summarized below.

7.4.1 Temporal Coordination

The dynamic aspects of visuomotor transformations and movement coordination have not been thoroughly considered in the current research. All steps from visual acquisition to object lifting are executed sequentially, following the data flow, and in the case of concurrent processes (such as in ventral and dorsal visual analysis), the adopted solution is simply that the faster process waits for the slower.

For what concerns visual cue integration, it seems that binocular cues are processed faster than monocular ones (Greenwald et al. 2005. This can be explained by the fact that interpretation of monocular cues normally requires semantic knowledge and hence some ventral stream processing, whilst stereoptic analysis is performed entirely within the dorsal stream. The consequence is that temporal constraints affect cue integration, and binocular cues are especially dominant during online action control.

Regarding coordination in action execution, it surely involves IPS and PMv, but other areas too. The supplementary motor area (Picard and Strick 2003; Ogawa et al. 2006) and the dorsal premotor cortex (Davare et al. 2006) are almost certainly part of the circuit dedicated to the sequencing of action components in reaching and grasping, but the most important role in action coordination is very likely played by the cerebellum (Doya 1999; Cotterill 2001; Ramnani et al. 2001). More complex mechanisms, and the contribution of other areas, seem to be required when grasping involves moving objects (Schenk et al. 2005; Sakata et al. 1997). An additional aspect that should be taken into account is the coordination between eye and arm movements (Johansson et al. 2001). In a research work largely complementary to that presented in this book, Chinellato et al. (2011) have developed a bio-inspired system for visuomotor integration based on the contextual learning and the coordinated control of eye gazing and arm reaching movements. This has allowed to produce a robotic system able to create its own internal model and its representation of the surrounding space by interacting with the environment directly, through a mutual adaptation of perception and action (Antonelli et al. 2014).

7.4.2 Tools

Although other primates can successfully learn how to use simple tools, humans are the only species specialized for tool manipulation. Indeed, most of our every-day activities require the handling of purposely manufactured objects. Tool use is thus an extremely revealing aspect for studying the evolution of the human brain as compared to monkeys'.

Familiarity with objects surely affect the way we interact with them (Gentilucci 2003), but tools seem to constitute a completely autonomous class of objects to this regard (Creem-Regehr and Lee 2005; Hoeren et al. 2013). It has been suggested that there are neural mechanisms especially dedicated to tool handling (Johnson-Frey et al. 2005; Orban and Caruana 2014), and that tools at reachable distances seem to constitute a focus of attention more powerful than other graspable objects (Handy et al. 2003). Sugioet al. (2003a) compared the fMRI activation of grasping familiar objects with explicit graspable features, such as handles, and other graspable objects without handles. The latter were found to elicit the usual AIP–PMv circuit, but the former apparently activated learned visuomotor associations at subcortical areas, like the rostral cingulate motor area. The same area is not activated for geometric objects with handle-like parts, implying that the activation is related to semantic meanings rather than to geometric features (Sugio et al. 2003b). According to the authors these results suggest the existence of a "direct" route from vision to action especial for tools, which is essentially affected by semantic and associative factors.

7.4.3 Grasping Force

The grasping literature distinguishes between grip and lift force, suggesting that, although related, they are controlled by parallel processes (Quaney et al. 2005), and that they are both independent from the reaching movement (Biegstraaten et al. 2006). The use of force has been much simplified in the proposed framework, but there are many factors which affects force distribution, and many of them are unpredictable before contact (Baud-Bovy and Soechting 2002). The influence of object weight, expected friction and compliance have already been mentioned, and the control of grasping forces, probably performed by AIP (Ehrsson et al. 2003), takes into account all of them (Gordonet al. 1993). Indeed, complex interactions between the streams seem to underlie the anticipatory scaling and the subsequent monitoring of grasping forces (Westwoodet al. 2000; Reichenbach et al. 2014). Nevertheless, the nature of the contact and the exact situation can be assessed only after touching the object, and two different control strategies for balancing forces and moments seem to be contextually pursued, and the resulting commands summed up (Zatsiorsky et al. 2004). The biggest challenge in modeling force control in grasping is probably the fast adaptation exhibited by subjects in learning the most appropriate force patterns in each condition. Salimi et al. (2000) suggest that this is achieved through the modulation of multiple internal representations, and Ulloa et al. (2003) implemented a plausible model of force learning based on cerebellar mechanisms.

7.4.4 Illusions

Visual and visuomotor illusions constitute interesting tools for studying complex perceptual mechanisms. For what concerns the two streams theory, common illusions have been found not to affect grip scaling as they do with size judgment (Servos et al. 2000). Force scaling seems to be also largely spared by illusions, but several factors such as relative size, delays and environmental conditions all affect the response to illusions (Westwood and Goodale 2003; Handlovsky et al. 2004). Some researchers argue that the observed effects are not due to the dualism between the dorsal and ventral streams (Franz et al. 2001; Dassonville and Bala 2004). Nevertheless, recent studies appositely designed to disambiguate the effect of illusions and solve the controversy, support the perception/action dualism (Kwok and Braddick 2003; Stöttinger and Perner 2006; Plewan et al. 2012).

As for brain damages, the modeling and simulation of illusory effects can be of great interest for the validation of modeling hypotheses on the function and connectivity of visual and visuomotor areas.

Other aspects related to vision based grasping and to the two streams theory that are worth further exploration and modeling efforts are: laterality and the task sharing between the left and right brain hemispheres (Cavina-Pratesi et al. 2006; Culham et al. 2006; Rice et al. 2007; Vainio et al. 2007); the controversial issue of visuomotor priming (Craighero et al. 2002; Cant et al. 2005; Yoon and Humphreys 2007) and the related problem of attention allocation (Craighero et al. 1999; Handy et al. 2003; Lavie 2005).

7.5 Conclusions

Interdisciplinary research involving high technology fields and life sciences is getting nowadays more and more common. Artificial intelligence has been from its very foundation a meeting-place for scientists of seemingly unrelated disciplines. Two such disciplines which met thanks to artificial intelligence are robotics and neuroscience, and although their encounter is producing very interesting developments, fundamental differences in research goals, methodologies and language prevent a more proficuous collaboration. To this regard, the main achievement of our work is to present in a real unified framework two approaches to the issue of vision-based grasping, neuroscientific and robotic, that are usually completely independent from each other.

References

Amedi A, Malach R, Hendler T, Peled S, Zohary E (2001) Visuo-haptic object-related activation in the ventral visual pathway. Nat Neurosci 4(3):324–330. doi:10.1038/85201

Antonelli M, Gibaldi A, Beuth F, Duran A, Canessa A, Chessa M, Solari F, del Pobil A, Hamker F, Chinellato E, Sabatini S (2014) A hierarchical system for a distributed representation of the peripersonal space of a humanoid robot. IEEE Trans Auton Mental Dev 6(4):259–273. doi:10.1109/TAMD.2014.2332875

Arbib MA, Billard A, Iacoboni M, Oztop E (2000) Synthetic brain imaging: grasping, mirror neurons and imitation. Neural Netw 13(8–9):975–997

Baud-Bovy G, Soechting JF (2002) Factors influencing variability in load forces in a tripod grasp. Exp Brain Res 143(1):57–66. doi:10.1007/s00221-001-0966-8

Begliomini C, Wall MB, Smith AT, Castiello U (2007) Differential cortical activity for precision and whole-hand visually guided grasping in humans. Eur J Neurosci 25(4):1245–1252. doi:10.1111/j.1460-9568.2007.05365.x

Biegstraaten M, Smeets JBJ, Brenner E (2006) The relation between force and movement when grasping an object with a precision grip. Exp Brain Res 171(3):347–357. doi:10.1007/s00221-005-0271-z

Binkofski F, Buccino G, Zilles K, Fink GR (2004) Supramodal representation of objects and actions in the human inferior temporal and ventral premotor cortex. Cortex 40(1):159–161

Bosco A, Breveglieri R, Chinellato E, Galletti C, Fattori P (2009) Influence of visual feedback on reaching activity in parietal area V6A. In: Annual meeting of the society for neuroscience, 2009

Cant JS, Westwood DA, Valyear KF, Goodale MA (2005) No evidence for visuomotor priming in a visually guided action task. Neuropsychologia 43(2):216–226. doi:10.1016/j.neuropsychologia.2004.11.008

Cavina-Pratesi C, Monaco S, McAdam T, Milner D, Schenk T, Culham JC (2007) Which aspects of hand-preshaping does human AIP compute during visually guided actions? Evidence from event-related fMRI. In: Annual meeting of the society for neuroscience

Cavina-Pratesi C, Valyear KF, Culham JC, Köhler S, Obhi SS, Marzi CA, Goodale MA (2006) Dissociating arbitrary stimulus-response mapping from movement planning during preparatory period: evidence from event-related functional magnetic resonance imaging. J Neurosci 26(10):2704–2713. doi:10.1523/JNEUROSCI.3176-05.2006

Chinellato E, Antonelli M, Grzyb B, del Pobil A (2011) Implicit sensorimotor mapping of the peripersonal space by gazing and reaching. IEEE Trans Autonom Mental Dev 3(1):43–53. doi:10.1109/TAMD.2011.2106781

Chinellato E, del Pobil AP (2008) fRI, functional robotic imaging: Visualizing a robot brain. In: IEEE international conference on distributed human-machine systems

Chinellato E, Grzyb BJ, Marzocchi N, Bosco A, Fattori P, del Pobil AP (2010) The dorso-medial visual stream: from neural activation to sensorimotor interaction. Neurocomputing (In Press). doi:10.1016/j.neucom.2010.07.029

Cloutman LL (2013) Interaction between dorsal and ventral processing streams: where, when and how? Brain Lang 127(2):251–263. doi:10.1016/j.bandl.2012.08.003

Cotterill RM (2001) Cooperation of the basal ganglia, cerebellum, sensory cerebrum and hippocampus: possible implications for cognition, consciousness, intelligence and creativity. Prog Neurobiol 64(1):1–33
Craighero L, Fadiga L, Rizzolatti G, Umiltà C (1999) Action for perception: a motor-visual attentional effect. J Exp Psychol Hum Percept Perform 25(6):1673–1692
Craighero L, Bello A, Fadiga L, Rizzolatti G (2002) Hand action preparation influences the responses to hand pictures. Neuropsychologia 40(5):492–502
Creem-Regehr SH, Lee JN (2005) Neural representations of graspable objects: are tools special? Cogn Brain Res 22(3):457–469. doi:10.1016/j.cogbrainres.2004.10.006
Culham JC, Cavina-Pratesi C, Singhal A (2006) The role of parietal cortex in visuomotor control: what have we learned from neuroimaging? Neuropsychologia 44(13):2668–2684. doi:10.1016/j.neuropsychologia.2005.11.003
Culham JC (2006) Functional neuroimaging: experimental design and analysis. In: Cabeza R, Kingstone A (eds) Handbook of functional neuroimaging of cognition. MIT Press, Cambridge, pp 53–82
Dassonville P, Bala JK (2004) Perception, action, and Roelofs effect: a mere illusion of dissociation. PLoS Biol 2(11):e364. doi:10.1371/journal.pbio.0020364
Davare M, Andres M, Cosnard G, Thonnard JL, Olivier E (2006) Dissociating the role of ventral and dorsal premotor cortex in precision grasping. J Neurosci 26(8):2260–2268. doi:10.1523/JNEUROSCI.3386-05.2006
Dearden AM, Demiris Y (2005) Learning forward models for robots. In: International joint conferences on artificial intelligence, pp 1440–1445
Doya K (1999) What are the computations of the cerebellum, the basal ganglia and the cerebral cortex? Neural Netw 12(7–8):961–974
Ehrsson HH, Fagergren A, Johansson RS, Forssberg H (2003) Evidence for the involvement of the posterior parietal cortex in coordination of fingertip forces for grasp stability in manipulation. J Neurophysiol 90(5):2978–2986. doi:10.1152/jn.00958.2002
Ehrsson HH, Fagergren A, Jonsson T, Westling G, Johansson RS, Forssberg H (2000) Cortical activity in precision- versus power-grip tasks: an fMRI study. J Neurophysiol 83(1):528–536
Ernst MO, Banks MS, Bülthoff HH (2000) Touch can change visual slant perception. Nat Neurosci 3(1):69–73. doi:10.1038/71140
Fattori P, Raos V, Breveglieri R, Bosco A, Marzocchi N, Galletti C (2010) The dorsomedial pathway is not just for reaching: grasping neurons in the medial parieto-occipital cortex of the macaque monkey. J Neurosci 30(1):342–3490. http://dx.doi.org/0
Franz VH, Fahle M, Bülthoff HH, Gegenfurtner KR (2001) Effects of visual illusions on grasping. J Exp Psychol Hum Percept Perform 27(5):1124–1144
Frey SH, Vinton D, Norlund R, Grafton ST (2005) Cortical topography of human anterior intraparietal cortex active during visually guided grasping. Cogn Brain Res 23(2–3):397–405. doi:10.1016/j.cogbrainres.2004.11.010
Fukuda H, Fukumura N, Katayama M, Uno Y (2000) Relation between object recognition and formation of hand shape: a computational approach to human grasping movements. Syst Comput Jpn 31(12):11–22
Gentilucci M (2003) Object familiarity affects finger shaping during grasping of fruit stalks. Exp Brain Res 149(3):395–400. doi:10.1007/s00221-003-1370-3
Goodale MA (2008) Action without perception in human vision. Cogn Neuropsychol 25(7–8):891–919. doi:10.1080/02643290801961984
Goodale MA, Milner AD (1992) Separate visual pathways for perception and action. Trends Neurosci 15(1):20–25
Gordon AM, Westling G, Cole KJ, Johansson RS (1993) Memory representations underlying motor commands used during manipulation of common and novel objects. J Neurophysiol 69(6):1789–1796
Greenwald HS, Knill DC, Saunders JA (2005) Integrating visual cues for motor control: a matter of time. Vis Res 45(15):1975–1989. doi:10.1016/j.visres.2005.01.025

Handlovsky I, Hansen S, Lee TD, Elliott D (2004) The Ebbinghaus illusion affects on-line movement control. Neurosci Lett 366(3):308–311. doi:10.1016/j.neulet.2004.05.056

Handy TC, Grafton ST, Shroff NM, Ketay S, Gazzaniga MS (2003) Graspable objects grab attention when the potential for action is recognized. Nat Neurosci 6(4):421–427, 431

Hoeren M, Kaller CP, Glauche V, Vry MS, Rijntjes M, Hamzei F, Weiller C (2013) Action semantics and movement characteristics engage distinct processing streams during the observation of tool use. Exp Brain Res 229(2):243–260. doi:10.1007/s00221-013-3610-5

Hu Y, Osu R, Okada M, Goodale MA, Kawato M (2005) A model of the coupling between grip aperture and hand transport during human prehension. Exp Brain Res 167(2):301–304. dio:10.1007/s00221-005-0111-1

Jeannerod M (1999) Visuomotor channels: their integration in goal-directed prehension. Hum Mov Sci 18(2):201–218, doi:10.1016/S0167-9457(99)00008-1

Jiang J, Shen Y, Neilson PD (2002) A simulation study of the degrees of freedom of movement in reaching and grasping. Hum Mov Sci 21(5–6):881–904

Johansson RS, Westling G, Bäckström A, Flanagan JR (2001) Eye-hand coordination in object manipulation. J Neurosci 21(17):6917–6932

Johnson-Frey SH, Newman-Norlund R, Grafton ST (2005) A distributed left hemisphere network active during planning of everyday tool use skills. Cerebral Cortex 15(6):681–695. doi:10.1093/cercor/bhh169

Kawato M (1999) Internal models for motor control and trajectory planning. Curr Opin Neurobiol 9:718–727

Kwok RM, Braddick OJ (2003) When does the Titchener Circles illusion exert an effect on grasping? Two- and three-dimensional targets. Neuropsychologia 41(8):932–940

Lavie N (2005) Distracted and confused?: selective attention under load. Trends Cogn Sci 9(2):75–82. doi:10.1016/j.tics.2004.12.004

Lim B, Ra S, Park F (2005) Movement primitives, principal component analysis, and the efficient generation of natural motions. In: IEEE international conference on robotics and automation, pp 4630–4635

Loftus A, Servos P, Goodale MA, Mendarozqueta N, Mon-Williams M (2004) When two eyes are better than one in prehension: monocular viewing and end-point variance. Exp Brain Res 158(3):317–327. doi:10.1007/s00221-004-1905-2

Metzinger T, Gallese V (2003) The emergence of a shared action ontology: building blocks for a theory. Conscious Cogn 12(4):549–571

Miall R (2003) Connecting mirror neurons and forward models. Neuroreport 14(16):1–3

Mon-Williams M, Tresilian JR (1999) Some recent studies on the extraretinal contribution to distance perception. Perception 28(2):167–181

Mon-Williams M, Tresilian JR (2001) A simple rule of thumb for elegant prehension. Curr Biol 11(13):1058–1061

Murata A, Gallese V, Luppino G, Kaseda M, Sakata H (2000) Selectivity for the shape, size, and orientation of objects for grasping in neurons of monkey parietal area AIP. J Neurophysiol 83(5):2580–2601

Nori F, Frezza R (2005) Control of a manipulator with a minimum number of motion primitives. In: IEEE international conference on robotics and automation, pp 2344–2349

Ogawa K, Inui T, Sugio T (2006) Separating brain regions involved in internally guided and visual feedback control of moving effectors: an event-related fMRI study. Neuroimage 32(4):1760–1770. doi:10.1016/j.neuroimage.2006.05.012

Orban GA, Caruana F (2014) The neural basis of human tool use. Front Psychol 5:310. doi:10.3389/fpsyg.2014.00310

Perry CJ, Tahiri A, Fallah M (2014) Feature integration within and across visual streams occurs at different visual processing stages. J Vis 14(2). doi:10.1167/14.2.10

Picard N, Strick PL (2003) Activation of the supplementary motor area (SMA) during performance of visually guided movements. Cereb Cortex 13(9):977–986

Plewan T, Weidner R, Eickhoff SB, Fink GR (2012) Ventral and dorsal stream interactions during the perception of the müller-lyer illusion: evidence derived from fmri and dynamic causal modeling. J Cogn Neurosci 24(10):2015–2029

Pouget A, Sejnowski TJ (2001) Simulating a lesion in a basis function model of spatial representations: comparison with hemineglect. Psychol Rev 108(3):653–673

Quaney BM, Nudo RJ, Cole KJ (2005) Can internal models of objects be utilized for different prehension tasks? J Neurophysiol 93(4):2021–2027. doi:10.1152/jn.00599.2004

Ramnani N, Toni I, Passingham RE, Haggard P (2001) The cerebellum and parietal cortex play a specific role in coordination: a PET study. Neuroimage 14(4):899–911. doi:10.1006/nimg.2001.0885

Reichenbach A, Thielscher A, Peer A, Bülthoff HH, Bresciani JP (2014) A key region in the human parietal cortex for processing proprioceptive hand feedback during reaching movements. Neuroimage 84:615–625. doi:10.1016/j.neuroimage.2013.09.024

Rice NJ, Tunik E, Cross ES, Grafton ST (2007) On-line grasp control is mediated by the contralateral hemisphere. Brain Res 1175:76–84. doi:10.1016/j.brainres.2007.08.009

Rizzolatti G, Luppino G (2001) The cortical motor system. Neuron 31(6):889–901

Roy A, Paulignan Y, Meunier M, Boussaoud D (2002) Prehension movements in the macaque monkey: Effects of object size and location. Mach Learn 88(3):1491–1499

Sakata H, Taira M, Kusunoki M, Murata A, Tanaka Y (1997) The TINS lecture. The parietal association cortex in depth perception and visual control of hand action. Trends Neurosci 20(8):350–357

Salimi I, Hollender I, Frazier W, Gordon AM (2000) Specificity of internal representations underlying grasping. J Neurophysiol 84(5):2390–2397

Salinas E, Sejnowski TJ (2001) Gain modulation in the central nervous system: where behavior, neurophysiology, and computation meet. Neuroscientist 7(5):430–440

Schenk T, Ellison A, Rice N, Milner AD (2005) The role of V5/MT+ in the control of catching movements: an rTMS study. Neuropsychologia 43(2):189–198. doi:10.1016/j.neuropsychologia.2004.11.006

Servos P, Carnahan H, Fedwick J (2000) The visuomotor system resists the horizontal-vertical illusion. J Mot Behav 32(4):400–404

Shadmehr R, Wise SP (2005) The computational neurobiology of reaching and pointing: a foundation for motor learning. MIT Press, Cambridge

Shikata E, Hamzei F, Glauche V, Koch M, Weiller C, Binkofski F, Büchel C (2003) Functional properties and interaction of the anterior and posterior intraparietal areas in humans. Eur J Neurosci 17(5):1105–1110

Shmuelof L, Zohary E (2005) Dissociation between ventral and dorsal fMRI activation during object and action recognition. Neuron 47(3):457–470. doi:10.1016/j.neuron.2005.06.034

Smeets JBJ, Brenner E, Biegstraaten M (2002) Independent control of the digits predicts an apparent hierarchy of visuomotor channels in grasping. Behav Brain Res 136(2):427–432

Stöttinger E, Perner J (2006) Dissociating size representation for action and for conscious judgment: Grasping visual illusions without apparent obstacles. Conscious Cogn 15(2):269–284. doi:10.1016/j.concog.2005.07.004

Sugio T, Inui T, Matsuo K, Matsuzawa M, Glover GH, Nakai T (1999) The role of the posterior parietal cortex in human object recognition: a functional magnetic resonance imaging study. Neurosci Lett 276(1):45–48

Sugio T, Ogawa K, Inui T (2003a) Multiple action representations of familiar objects with handles: an fMRI study. In: European conference on visual perception

Sugio T, Ogawa K, Inui T (2003b) Neural correlates of semantic effects on grasping familiar objects. Neuroreport 14(18):2297–2301. doi:10.1097/01.wnr.0000092474.09492.3a

Talairach J, Tournoux P (1988) Co-planar stereotaxic atlas of the human brain. Thieme, Stuttgart

Tankus A, Fried I (2012) Visuomotor coordination and motor representation by human temporal lobe neurons. J Cogn Neurosci 24(3):600–610

Tanné-Gariépy J, Rouiller EM, Boussaoud D (2002) Parietal inputs to dorsal versus ventral premotor areas in the macaque monkey: evidence for largely segregated visuomotor pathways. Exp Brain Res 145(1):91–103. doi:10.1007/s00221-002-1078-9

Ulloa A, Bullock D (2003) A neural network simulating human reach-grasp coordination by continuous updating of vector positioning commands. Neural Netw 16(8):1141–1160. doi:10.1016/S0893-6080(03)00079-0

Ulloa A, Bullock D, Rhodes BJ (2003) Adaptive force generation for precision-grip lifting by a spectral timing model of the cerebellum. Neural Netw 16(5–6):521–528. doi:10.1016/S0893-6080(03)00094-7

Vainio L, Ellis R, Tucker M, Symes E (2007) Local and global affordances and manual planning. Exp Brain Res 179(4):583–594. doi:10.1007/s00221-006-0813-z

Valyear KF, CavinaspsPratesi C, Stiglick AJ, Culham JC (2007) Does tool-related fMRI activity within the intraparietal sulcus reflect the plan to grasp? Neuroimage 36(Suppl 2):T94–T108. doi:10.1016/j.neuroimage.2007.03.031

van de Kamp C, Zaal FTJM (2007) Prehension is really reaching and grasping. Exp Brain Res 182(1):27–34. doi:10.1007/s00221-007-0968-2

Verhoef BE, Vogels R, Janssen P (2011) Synchronization between the end stages of the dorsal and the ventral visual stream. J Neurophysiol 105(5):2030–2042. doi:10.1152/jn.00924.2010

Westwood DA, Dubrowski A, Carnahan H, Roy EA (2000) The effect of illusory size on force production when grasping objects. Exp Brain Res 135(4):535–543

Westwood DA, Goodale MA (2003) Perceptual illusion and the real-time control of action. Spat Vis 16(3–4):243–254

Wokke ME, Scholte HS, Lamme VAF (2014) Opposing dorsal/ventral stream dynamics during figure-ground segregation. J Cogn Neurosci 26(2):365–379

Wolpert DM, Ghahramani Z (2000) Computational principles of movement neuroscience. Nat Neurosci 3(Suppl):1212–1217. doi:10.1038/81497

Yoon EY, Humphreys GW (2007) Dissociative effects of viewpoint and semantic priming on action and semantic decisions: evidence for dual routes to action from vision. Q J Exp Psychol: Colchester 60(4):601–623. doi:10.1080/17470210600701007

Zanon M, Busan P, Monti F, Pizzolato G, Battaglini PP (2010) Cortical connections between dorsal and ventral visual streams in humans: evidence by tms/eeg co-registration. Brain Topogr 22(4):307–317. doi:10.1007/s10548-009-0103-8

Zatsiorsky VM, Latash ML, Gao F, Shim JK (2004) The principle of superposition in human prehension. Robotica 22(2):231–234. doi:10.1017/S0263574703005344

MIX
Papier aus verantwortungsvollen Quellen
Paper from responsible sources
FSC® C105338

If you have any concerns about our products,
you can contact us on
ProductSafety@springernature.com

In case Publisher is established outside the EU,
the EU authorized representative is:
Springer Nature Customer Service Center GmbH
Europaplatz 3, 69115 Heidelberg, Germany

Printed by Libri Plureos GmbH
in Hamburg, Germany